The Contemporary "Origins" Debate

A Short Primer and Creation Model Proposal

Neo-Darwinian Evolution, Intelligent Design and Creation.
Is a peaceful coexistence possible?

J. P. Hubert Jr. BA (Biology) MD FACS

ISBN: 978-1-4303-2208-5

Preface

The "arrow" or trajectory of scientific discovery in the life sciences is undoubtedly all in one direction that is, toward progressively increasing layers of organization and complexity not just in space but also in time; particularly in molecular biology where some biochemical reactions involving dozens of intricate steps occur in an extraordinarily short time interval.

The question for many serious students and interested laymen is whether the Neo-Darwinian mutation/selection mechanism [of natural selection acting on random (chance driven) germ cell mutational change] is powerful enough to completely explain the incredibly diverse panoply of life on earth. Discoveries continually being made in the modern sciences of cosmology, biochemistry, physics, paleontology etc. have brought this heretofore almost universally held (by biologists) assumption into serious dispute. It is in this context that some have raised the question of whether in addition to chance and necessity, "design" as mechanism has also played a role. Proponents of this view have posited what has become known as Intelligent Design (ID) theory which holds that not all life on earth can be attributed to the mutation/selection mechanism but that certain structures, biochemical pathways etc. are the result of either information or structures appearing in the fossil record in completed or final form from which minor variation (micro-evolutionary changes only) has occurred. ID proponents allege that criteria exist for determining these alleged occurrences. Most Neo-Darwinists dispute their assertions.

Unfortunately, the Origin's question is fraught with misunderstanding as a result of much imprecision with respect to the use of terminology as well as the constant mixing of theological and philosophical concepts with those of a strictly natural scientific nature. For example, the question of whether anything other than the material realm (which science studies) exists is fundamentally a philosophical one which can also be addressed by theology but not by natural science limited as it is to the study of the material realm. The overarching construct which some Neo-Darwinists have promulgated (concomitant with their study of the mutation/natural selection mechanism) is really a metaphysical system or total worldview which can be referred to as philosophical naturalism (PN) or radical Darwinism. No such ontological development is either necessary or appropriate based on the convention of modern science (methodological naturalism [MN]) limited as it is to the investigation of material phenomena only. Radical Darwinists who insist upon the conflation of Neo-Darwinism as mechanism with the much more encompassing philosophical naturalism or materialist philosophy do it a disservice and in so doing demonstrate the worst kind of example for those who possess a genuine desire to search for the truth and to come to a better understanding of the limits of modern biological science where investigation of "Origin's" is concerned. The present work is intended to shed some much needed light on this intensely debated topic.

Contents

6

Introduction

The "Origins" debate has raged since Darwin's publication of *On the Origin of Species* in 1859.[1] In the wake of the Enlightenment and Francis Bacon's advocacy for what has become known as "methodological naturalism" in science[2], a widening gulf has grown between the relevant disciplines of science, philosophy and theology.[3] Today, many scholars are completely ignorant of the foundational principles of other disciplines as well as the increasingly complex nature of their detailed subject matter. This is particularly true in the sciences, has contributed to a breakdown in cross-disciplinary understanding and has allowed each to engage in "mission creep" or "boundary creep."[4] Too often, scientists

---

[1] Charles Darwin. *On the Origin of Species*, London: John Murray, 1859, in which the term "Origins" was used in the context of explaining the derivation of living organisms. Later the term became associated with the origin of the cosmos as well.

[2] Francis Bacon [1605]. *The Advancement of Learning*, (Oxford: Clarendon Press, 1868) in which formal and final causality were labeled metaphysics and effectively elimated from modern science and in which Aristotle's efficient causality was redefined by David Hume as event (psychological) causation experienced as sense impressions rather than object [things] where there is an ontologically necessary or sufficient causality associated with a given effect. As a result, Hume also rejected substance in the classical sense. See William A. Wallace *The Modeling of Nature*: Philosophy of Science and Philosophy of Nature in Synthesis. (Washington D.C.: Catholic University of America Press, 1996), 258-262.

[3] The problem is multi-factorial and includes the tremendous increase in knowledge which has occurred in each discipline and sub-discipline, the lack of proper preparation in philosophy (particularly the classical Greek and Roman thinkers) by those who undertake scientific careers including the lack of understanding of first principles of being and the rules of logic, demonstration, dialectic, rhetoric and their proper use. Universities are increasingly compartmentalized and isolated, rather than being centers of shared knowledge across disciplines.

[4] Not understanding the foundations and boundaries of their own discipline or that of others, many engage in developing a full blown "worldview" or metaphysical construct out of their own limited vantage point. A case in point is Tippler's *The Physics of Immortality*, (1994) in which he dedicates an entire book to establishing a mechanism for universal resurrection by the use of his knowledge of general relativity and astrophysics. See bibliography for details of citation. An adequate knowledge of philosophy

"wander" into a discussion of metaphysical and theological questions despite claiming to limit themselves to *methodological naturalism*[5] and the natural realm alone.[6] This tendency betrays at a minimum, a lack of classical preparation in philosophy and a not so hidden preference for philosophical naturalism (full blown materialism) which manifests itself as scientism.[7] For example, it is well known that the vast majority of academic biological scientists ascribe to both methodological and philosophical naturalism. It is clear that many either do not appreciate the difference between the two, or intentionally choose to blur those distinctions in order to privilege their own overall worldview (philosophical naturalism) with the patina of scientific respectability.[8] This is despite chastising critics of Darwinism, including creationists and ID theorists[9] for "contaminating" science with theology and metaphysics. Given their own disregard for the rules of science, (as Darwinists have promulgated and enforced them in the form of methodological naturalism), one can hardly blame Darwinian critics for responding in kind. Despite a concerted effort by ID theorists to obtain a hearing for their

---

(particularly the Aristotelian/Thomistic synthesis would have prevented this futile exercise in subsuming theology and philosophy into physics.

[5] The question of what belongs in the realm of science is actually a philosophical not scientific question.

[6] No doubt this is primarily because in the post-modern West, "science" is seen as the only source of actual (truth) knowledge and all topics benefit from a "scientific" association. This is tragic since scientific knowledge is often less secure (due to its inductive nature) than is knowledge of a deductive type which is associated with higher mathematics, logic and even higher metaphysics. Examples of scientists who have extended their scientific theorizing into metaphysics and theology include; Frank Tippler, Richard Dawkins, Douglas Futuyma, Niles Eldridge, Ernst Mayr, the deceased Carl Sagan and Stephen Gould and many others. See the bibliography; Wiker and Percey for a discussion of "science" as the only source of truth in the contemporary West.

[7] In which it is assumed that reality is limited to the material realm.

[8] The list of offenders is so long as to defy any short reference list but includes some of the most famous Darwinian biologists of the 20th and 21st centuries.

[9] Intelligent Design is a relatively new attempt to legitimize "design" as part of science which centers on the recognition of specified complexity, identified by the use of statistics and information theory. See references in the bibliography by Dembski primarily, but also Meyer, Johnson, and Behe.

research program, Darwinists continue to insist upon methodological naturalism as the only proper way in which to do science and remain totally opposed to ID in principle.[10] This is perplexing since Darwinists frequently leap with both feet into clearly metaphysical and theological territory. A review of any contemporary college or graduate level biology text will easily confirm this assertion.[11]

The fact is that scientists cannot have it both ways. If they wish to rigidly enforce methodological naturalism as the only valid way to do *origin* science as well as *operation* science[12], then they must stop resorting to metaphysical and theological arguments and comments altogether. This means that many of their books and other publications will need to be re-written or severely edited. To do otherwise is simply dishonest. Until they do so, they have no right to criticize those opponents of Darwinism and philosophical naturalism who respond in kind with metaphysical or theological arguments and replies. It is disingenuous to argue that skeptics are not engaging in "scientific" critiques when Darwinists invite them by injecting metaphysics and theology into their own theories and commentaries. The playing field must be fair and level and all must agree to the rules of engagement in advance if the science portion of "Origins" is to advance. If that is impossible, then methodological naturalists will simply have to live with the re-insertion of design back into science, a situation ultimately of their

---

[10] The history of methodological naturalism in science is roughly 400 years old, originating with Francis Bacon. Prior to that time, design was not excluded (in principle) from science. Methodological naturalism is clearly the one contemporary generally agreed upon way in which to do science but not the only way as design theorists have argued.

[11] See Douglas Futuyma's textbook for example which is rife with metaphysical language and concepts none of which can be established if a strict adherence to methodological naturalism is followed.

[12] See J. P. Moreland's discussion for a detailed accounting of the differences in operation vs.: origins science.

own making and one which is worthy and amenable to testing.[13]

---

[13] Note that many Darwinists (e.g. Eugenie Scott) contend that design is untestable and yet also argue that it is not part of science in principle. Obviously if design is ruled-out "in-principle" that is categorically, it can not be tested. If it is allowed as one possible explanation for the development of life on earth, then it is testable. Chance, law and design alone or in combination are all testable by arguing to the best explanation or by an appeal to the greatest explanatory power.

2. Nature of the Debate

The "Origins" debate is an instance where the basic question is metaphysical, not fundamentally scientific even though a great deal of modern science is without doubt relevant. This explains why there has been so much animus on both sides. Each sees their entire *worldview* riding on the outcome of the so-called evolution/creation controversy. This is sad, since neither side will have its worldview vindicated should they prevail.[14] The reason is simply that irrespective of how the universe and life actually developed, whether it be by chance, necessity (law), design or some combination thereof,[15] all three are necessarily dependent on an immaterial first cause and this is knowable on the basis of a proper understanding of the philosophy of nature and metaphysics. That is to say, even if it should be established that life arose "naturally" from non-life (abiogenesis) over 3.5 billion year ago (an extremely unlikely scenario), and that the neo-Darwinian synthesis alone is completely capable of explaining all subsequent development of life on earth including humans, it would not eliminate the need for an ultimate immaterial first cause of the material universe and thus life itself.[16] The characteristics of the immaterial first cause or unmoved mover are further "fleshed-out" by a consideration of the data from science and Divine Revelation, but an acceptance of the data from revelation is not required in

---

[14] For example, even if ID can establish that design can be detected in biological objects, it does not prove that they were caused by an immaterial first cause, (an unembodied designer as Dembski calls it) a point that ID theorists repeatedly raise in arguing to have ID admitted into contemporary science.

[15] All of which are possible techniques, ways or mechanisms by which the universe and life could theoretically have developed either independently or in combination.

[16] This is a point that those properly educated in philosophy readily understand and for those who strictly adhere to methodological naturalism alone, it is easily appreciable that science as presently formulated cannot refute such a claim since according to methodological naturalism, science involves primarily material but not classical efficient, formal or final causality.

order to establish the necessity of an immaterial first cause.

Philosophy, (specifically a detailed study of the philosophy of nature) is adequate for revealing the necessity of an immaterial first cause of material reality, the so-called "first unmoved mover" of Aristotle.[17] Science on the other hand is incapable of refuting the existence of God in principle and while it can provide information about the nature and characteristics of God it is unable to establish God's existence in isolation.[18] The Design argument and the modern Kalam Cosmological argument popularized by Craig[19], both of which employ modern science are fundamentally philosophical arguments not scientific ones. The Greek (Aristotle for example) pagans were perfectly capable of recognizing that the material realm could not be its own first cause and this has not changed in over 2000 years despite the advent of incredible amounts of scientific knowledge.[20] Thus critics of *creationism* and *intelligent design* if intellectually honest should admit that to posit an immaterial first cause of material reality is common to both pagan and theistic thinkers and thus the argument stands as a legitimate intellectual (philosophical) position separate from religious attachments. Radical Darwinists (philosophical naturalists) are guilty of committing the genetic fallacy when they impugn the motives of those who argue for design in biology whether it be "front-loaded" or specially inserted "along side"

---

[17] See William A Wallace. *The Elements of Philosophy*: A Compendium for Philosophers and Theologians. (New York: Alba House, 1977), 55-58. See also Aristotle's Physics, Bk. I, ch 9 (430a 23-24). For a contemporary Thomistic account of the Cosmological argument see; Bruce R. Reichenbach. *The Cosmological Argument*: A Reassessment. (Springfield, Ill.: Charles C. Thomas Publishers), 1972.

[18] Whether adhering to methodological naturalism or not, science cannot directly study final causality or immaterial reality. At best it can make inferences about it.

[19] William Lane Craig. *Reasonable Faith*: Christian Truth and Apologetics. (Wheaton, Ill: Crossway Books, 1984), 91-122

[20] See Aristotle's Posterior Analytics and the conclusion of his Physics, Bk. I, ch. 9 (regarding the first unmoved mover).

natural processes as ID theorists propose.[21]

Thus far then, we have indicated that the elite contemporary science community insists upon limiting science to *methodological naturalism* both in standard *operation* science and in *origins* science despite frequent forays into metaphysics and theology in which they frequently attempt to establish their overall worldview (philosophical naturalism).[22] This explains at least one motivation behind their refusal to consider "design" as a possible mechanism by which life began on earth and subsequently developed. We have indicated that if the contemporary science community refuses to desist from this practice, then ID theorists and other critics of neo-Darwinism and philosophical naturalism are justified in countering with metaphysical and religious arguments of their own and by continuing to propose and test *design* as a potentially viable concept/mechanism in science. We have also seen that none of the "3" mechanisms (chance, necessity or design [in biology]) are adequate alone or in combination with respect to disproving the existence of an immaterial (transcendent) first cause or unmoved mover of the universe, since all "3" are ultimately dependent upon a transcendent (immaterial and extra-universal) cause for the existence of the universe and derivatively life itself.

---

[21] Peter, Kreeft. *Socratic Logic:* A Logic Text Using Socratic Method, Platonic Questions, and Aristotelian Principles. (South Bend, Indiana: St. Augustine Press, 2004) 81.

[22] Richard Dawkins is perhaps the best but not only representative of this tendency.

14

3. What of "Creationism"

At this point a word should be said about what has been termed "Creationism" also referred to as *creation science*. The primary characteristic which defines "creation science" is a *literalist or fundamentalist Biblicist* interpretation of scripture particularly Genesis 1 & 2 and Exodus 20: 11. According to this view, God created the entire material realm in six 24 hour days in the recent past, (6-12 thousand years ago [young earth]) not millions or billions of years ago [the so-called "old earth" view]. Young Earth Creationism specifically rejects standard inflationary "Big Bang" cosmology as well as disputing the standard geological (uniformitarian) record, both of which suggest a 4.5 billion year old earth. Creation science advocates catastrophism in the recent past in the form of a truly world-wide flood, alleges that much of contemporary science is in error including the assumption of a constant speed of light, and a universe which is billions of light years in size and billions of years old. The creationist (Creation science) view arises out of one very specific interpretation of scripture in which the word "yom" (the Hebrew word for day) means one 24 hour period. While it clearly can have that meaning, *yom* can also refer to long periods or epochs of time, (such as "in the day" of the Emperor Julius Caesar or the Roman Empire) or the period from sundown to sunup (12 hours). While the creationist view could theoretically be true, it does require abandoning virtually all of modern cosmology, physics, and biology as false and untrustable.[23] While all of these disciplines may indeed be

---

[23] Such an assumption virtually assumes that man in his "fallen state" is incapable of ascertaining the truth about nature without specific guidance in the form of Divine Revelation a belief which is apparently grounded in a particular interpretation of Calvin's TULIP doctrine. On careful analysis this position is untenable. Many discoveries made by secular scientists have been confirmed by Divine Revelation such as the dietary prescriptions given to the ancient Israelites found in the Old Testament, the scientific validity of which was only recently confirmed, long after the same data was originally presented in scripture. These scientific discoveries were made without reference or

wrong, it seems <u>extremely</u> unlikely.[24]

An explanation which allows one to harmonize the data from theology, science and philosophy is that the references to "days" in Genesis and Exodus are not meant to be taken as 24 hour days. A detailed discussion of the issue is beyond the scope of this work however.[25] Henceforth, when the phrase; *special creation* is alluded to, it <u>will not</u> imply 24 hour periods of time or the "young earth" position, but rather "infusions" of either information and or new biological *forms* into the natural realm in whatever temporal relationship is defined e.g. usually thousands or millions of years of standard geological time. The author assumes that the earth is billions not thousands of years old for the purposes of this work and the "Origins" model which is proposed herein.

This writer is aware that some Christians[26] who hold a high view of Sacred Scripture contend that anything other than a literal fundamentalist Biblicist view of Genesis (6, 24 hour creation days roughly six thousand years ago) is incompatible with the first 1850 years of Christian exegesis and the views of some orthodox Jewish rabbinical sources which they purport invalidates the view posited in this work. The "Origins" model proposed herein would be capable of empirically testing their claims as well as those of ID and DE proponents and if the data increasingly support them, the

---

regard to scriptural claims. The creationist view is also completely dependent upon a strict literalist or biblical fundamentalist interpretation of relevant scripture which represents only one of a number of possible or plausible interpretations.

[24] Some have posited a conspiracy by atheist scientists who wish to discredit scripture, particularly Christianity, but there are many theists especially Christians among the science community who enthusiastically embrace the standard geological/astrophysical time-table and who are able to harmonize it with an orthodox rendering of sacred scripture.

[25] For a discussion of the "days" of creation from 3 different perspectives, see: *The Genesis Debate*: Three Views on the Days of Creation. Edited by David G. Hagopian, (Mission Viejo, California: Crux Press, Inc. 2001).

[26] Both Roman Catholics and Evangelical Protestants of a more Fundamentalist stripe. A detailed treatment of their position is beyond the scope of this work.

author would be willing to reconsider his views.

18

4. The Contemporary Understanding of Science

In this primer the author will attempt to place the contemporary "Origins" context in greater perspective by articulating some of the relevant yet distinct contributions made by science, philosophy and theology. It also proposes a testable "creation model" (in which both neo-Darwinian mechanisms and Creation/Design play a part) with which to compare the standard neo-Darwinian paradigm (naturalistic evolutionary model only) in explaining the origin and subsequent development of life on earth. In that regard it is noteworthy that several new books appeared in 2004 which added fuel to the already contentious "Origins" debate.[27] Unfortunately, most of the discussion on "Origins" including these newer works, whether with respect to the origin of the cosmos or life on earth is approached primarily from a strictly scientific perspective. In reality, the "Origins" question is fundamentally but not solely philosophical that is, metaphysical as outlined above. The ultimate question in "Origins" is whether material reality even at the universal level can be its own first cause.[28] Classical metaphysics and an understanding of the philosophy of nature (even without specific input from theology which is in complete agreement with the classical metaphysical position) updated in light of contemporary scientific

---

[27] William A. Dembski and Michael Ruse. *Debating Design*: From Darwin to DNA. (Cambridge: Cambridge University Press, 2004); Guillermo Gonzalez and Jay W. Richards. *The Privileged Planet*: How Our Place in the Cosmos is Designed for Discovery. (Washington, D.C. Regnery, 2004); William A. Dembski, *Uncommon Dissent*: Intellectuals who Find Darwinism Unconvincing. Wilmington Delaware: (ISI Books, 2004). Many important and related issues are discussed in each of these books.
[28] This question is what fuels the "Origins" debate and makes it extremely contentious. Philosophical naturalists appear to utilize science to establish that there is no immaterial reality while holding that science is limited only to methodological naturalism. This is intellectually dishonest since naturalists of this ilk are waging a primarily metaphysical argument without proper metaphysical resources. Such a technique is nothing but sophistry.

understanding, answers a resounding no to that question[29].

    As alluded to above, since the time of Francis Bacon modern science has embraced *methodological naturalism* as the acceptable way of engaging in empirical investigation and it has been remarkably successful (operationally) in advancing overall scientific knowledge and related technology. As a result, all metaphysical questions which Aristotle and Aquinas classified as being associated with formal and final causality are ruled out of bounds in principle, thus limiting science to the investigation of the natural realm from the perspective of material causality.[30] Questions that concern ultimate purpose, goals, values, design

---

[29] Aristotle, Aquinas and William A. Wallace. *The Elements of Philosophy*: A Compendium for Philosophers and Theologians (New York: Alba House, 1977), 55-58, 123-127, & 140. See also William A. Wallace. *The Modeling of Nature*: Philosophy of Science and Philosophy of Nature in Synthesis. (Washington D.C.: Catholic University of America Press, 1996). This immaterial first cause or unmoved mover is ultimately necessary in order to explain the origin of the highly specified and complex material world which natural science investigates. Any "being" (existence) that the universe or its contents enjoy is derivative. Only the immaterial first cause (God) is being as being, or being personified. A proper study of the philosophy of nature reveals that such an entity is required without specifically delineating its attributes which is left for the philosophy of religion, higher level metaphysics and theology.

[30] Although with respect to neo-Darwinism it is very questionable whether the mutation/natural selection mechanism is capable of providing "efficient" causality in addition. Darwinism is steeped in Humean logical empiricism and thus faces many problems including an appeal to "causation" in which a strictly psychological rather than ontological link is posited between two events. This reduces cause/effect relationships to temporal priority and constant conjunction of "events" rather than "things." This is in contradistinction to an Aristotelian view in which causality is understood as a true ontological relationship of things either necessarily or sufficiently connected (referred to classically as efficient causality). Hume's view follows as a consequence of his skepticism and the influence of Kant and Hegel's idealism. See Wallace, *The Modeling of Nature*, 258-262. The issue of whether the neo-Darwinian paradigm is capable of providing an adequate explanation for all life on earth is one not only of material but more importantly efficient causality as understood in the classical Aristotelian formulation, and one which is inadequately addressed by the Humean inadequate substitute of "causation" in this author's estimation. This issue is also the locus of significant on-going debate between Darwinian proponents and skeptics whether presented in those terms or not. Again the debate is founded in a disagreement about underlying philosophical assumptions.

parameters, even optimally functional physiological states etc. technically have no place in the discussion of either the origin of the cosmos or life itself (assuming methodological naturalism) since they involve questions of formal and final causality in Aristotelian/Thomistic terms.[31] Thus the many misguided statements from physical and biological scientists who refer to metaphysical questions as if they were scientific ones are most disconcerting, serving only to further "muddy" the waters.[32]

Suffice it to say that no discussion of "Origins" either of the cosmos or life itself can proceed without recognizing that the issue involves at least "3" disciplines; philosophy, science and theology. This is particularly important since Western science as mentioned above, is limited by convention to *methodological naturalism* while many scientists demonstrate a virtual lack of philosophical and theological understanding based on their repeated and unsuccessful forays into metaphysics and theology.[33] Given the frequent metaphysical comments of premiere academic scientists, it no doubt comes as a surprise to many aspiring scientists that **science is not synonymous with philosophical naturalism or**

---

[31] Thus according to methodological naturalism; physics, chemistry and biology do not involve questions of purpose, design, goals etc. despite the fact that no one can learn these disciplines without resorting to teleological models and thinking. This is certainly the case in practice, particularly in the practical and original research sciences.

[32] Richard Dawkins for example entitled an entire book in refutation of design at the level of the universe and life itself and has further written that Darwin made it possible to be an intellectually fulfilled atheist. These are clearly metaphysical comments/issues and conclusions which can not be drawn from the data of neo-Darwinism even if one assumes that the standard synthesis is completely capable of explaining the development of all life on earth. See his: The *Blind Watchmaker*: Why the Evidence of Evolution Reveals a Universe without Design. New York: W.W. Norton & Company, 1996. Note also that staunch Darwinist Kenneth Miller makes a similar point in his book *Finding Darwin's God*. A Scientist's Search for Common Ground Between God and Evolution. (New York: Harper Collins 2002), 184 -185 for example (Perennial Edition), in which he impugns the likes of Dawkins for injecting metaphysics into biology.

[33] Examples of contemporary physical and biological scientists who (based on their writings) are virtually ignorant of philosophy and theology are too numerous to mention in contradistinction to those of an earlier era, e.g. Newton, Galileo etc many of whom were well versed.

atheism; a metaphysical position which claims that reality is completely limited to the material realm and thus that there is no God, heaven or life after death.[34] According to philosophical naturalism, no immaterial (spiritual) reality exists, by definition. While this position could (theoretically) be true, it is fundamentally unprovable by science as presently formulated and is metaphysically unfounded as indicated above. Yet as strange as it may seem, philosophical naturalism enjoys great popularity among many academics despite enjoying no privileged or even equivalent metaphysical position over its most commonly held rival (the claim that reality is composed of both the material and immaterial). The latter of course is a position which is common to all "3" of the monotheistic religions and the most widely held both now and throughout the past 2000+ years. It was also held by the great Greek pagans Socrates, Plato and Aristotle as well as virtually all of the founders of modern science.

If the discussion of "Origins" is to include questions of purpose, goals, meaning, etc. it must be admitted that said discussions have extended beyond what science (assuming methodological naturalism) alone is capable of providing. The most that can be said for example about the so-called standard hot inflationary model of "Big Bang" cosmology is that at present it appears to call for a finite universe in time and space the origination of which has no explanation in the material realm.[35] This is compatible with the monotheistic claim that a transcendent (extra-universal) entity or being of infinite intellect and power "created" the universe in some way. To make such a claim

---

[34] Many methodological naturalists particularly biologists have allowed their methodological naturalism to "seep over" into philosophical naturalism without articulating the difference such as, Dawkins, Dennett, Futuyma, Lewontin, Wilson etc. This undermines their insistence that science should be limited to methodological naturalism which cannot by definition entertain questions of purpose, meaning, planning or goals. Their criticism of creationists and ID theorists thus rings hollow.

[35] It is presently impossible to know anything about the universe prior to about 10(-34) seconds after the "Big Bang" according to astrophysicists.

however, is strictly speaking a philosophical and religious rather than scientific one irrespective of how "good a fit" it might presently appear to be. Non-theists however are entitled to claim that the data are also compatible with a mystery, despite the fact that to do so would seem to argue from an inferior intellectual position. To invoke *multiple universes* or *imaginary time* in order to escape the metaphysical implications as some have done belies a degree of desperation and a clear incursion into metaphysics and away from science as defined by methodological naturalism.[36] If the history of science is any guide however, one should not assume that the present cosmic model will be the final one and the situation could certainly change.

---

[36] Stephen Hawking. *A Brief History of Time*: From the Big Bang to Black Holes. (New York: Bantam Books, 1988) [Paper back edition], 134-139 for example. Note also J. Leslie. *Universes*. (London: Routledge, 1989). For a refutation of Hawking's imaginary time see Craig's *Reasonable Faith*: Christian Truth and Apologetics. (Wheaton, Ill: Crossway Books, 1984).

5. Origin's Models Considered

While the author's general view of "Origins" has changed many times over the past 35 years, it seems to be stabilizing along the lines of what Dr. Walter Bradley proposes in his chapter from the recent book, *Debating Design*.[37] He argues that the "fine-tuning" of the physical constants[38] and the apparent irreducible complexity of basic biopolymers[39] such as DNA and RNA, particularly in the context of the origin of life dilemma[40] are the best overall evidence that both were *designed* and by implication generated by an unimaginably superior intellect.[41] His use of the *design* concept is fairly applicable to the way many contemporary ID theorists use it despite the fact that some design theorists are apparently unwilling to specifically detail which organisms/organs/systems were designed and when they first appeared.[42]

Bradley's conclusions represent a modification of the standard teleological argument, one of the classical arguments from philosophy for the existence of God, updated with

---

[37] See his chapter in *Debating Design*, 331-351.

[38] Which are "just-right" for life.

[39] Bradley, 350.

[40] To date no strictly material (inorganic to organic) explanation for first-life is forthcoming (abiogenesis).

[41] Bradley is a well-known Christian theist and scientist who has written prolifically on "Origins."

[42] See the two books by Dembski and note that he accepts Behe's thoughts regarding intelligently designed (irreducibly complex) entities such as the bacterial flagellum, the clotting cascade and the cilium as outlined in *Darwin's Black Box*. An exception is Stephen Meyer, *Debating Design*, (371-391) in which he argued that first life and the Cambrian information "explosion" are examples of intelligent design and in which he specifically follows the standard "Old Earth" timetable. Thus his predictions are specific to time and type of special insertion (infusion) of ID. See Dembski's *No Free Lunch* for an extensive discussion re: whether ID as a research program should be required to provide testability and predictability, 311-379.

contemporary findings from both the physical and biological sciences. With respect to the specific details of the "Origins" of the cosmos and "first-life" one assumes that Bradley advocates special creationism (2 episodes of specially inserted Intelligent Design, i.e. the creation of the universe [being] and the creation of first life) rather than theistic evolution although he declined to indicate whether his view calls for only two or multiple interventions at various stages in the history of the universe and the earth (ala the progressive special creationism of Hugh Ross et al.) or whether the "design" was completely "front-loaded" into the cosmos through the incredible fine-tuning of physical constants (ala the theistic evolution of Fr. Stanley Jaki or Kenneth Miller), which the author assumes is <u>not</u> his position. Note that both scenarios can accommodate undetectable manipulation of environmental conditions including weather systems, other astro/geo/physical phenomena etc. (existing alongside either one), in the form of contemporary neo-Darwinian mutation/natural selection.[43] In theory, (and leaving any scriptural input aside for the moment) it would seem that each of these two basic "Origins" scenarios is possible and that in principle neither is mutually exclusive that is to say both could have been involved in the origin of life and its further development. Such a situation might appear to violate Occam's razor which states simply that one must not multiply causal entities unnecessarily, but then this is true of any scenario which appeals to more than chance, law or design alone as causal mechanisms. In fact no one in the contemporary origin of life debate seems to favor chance, law, or design exclusively to the total exclusion of the others.[44] Even committed atheist/agnostic radical neo-Darwinists (philosophical naturalists) concede that

---

[43] In which virtually all "Origin of Life" scholars agree that genetic mutation and natural selection occurs at the species level (at a minimum). The debate occurs with respect to how much further in geological time the process can be extrapolated to higher taxonomic levels such as genus, family, class, order, phyla and kingdom.

[44] Darwinists for example invoke both chance and natural selection (i.e. law or necessity); special creationists who invoke design also invoke minor variation and adaptation in the form of genetic mutation and natural selection (chance and necessity as well as design).

both chance and necessity (law in the form of natural selection at least) are involved.

Judeo-Christian Scripture[45] however does seem to demand that both were operative that is, initial condition setting including the fundamental constants of physics (which theists are justified in imagining as a form of fiat ex-Nihilo special and progressive creation[46] of the universe from nothing by a transcendent and omnipotent God, and subsequent intermittent supernatural interventions [by God] on several additional occasions).[47] Hugh Ross and Fazale Rana (Reasons to Believe) seem to favor both although they reject the term "theistic evolution" preferring to call (the literally thousands or millions of separate creative events called for by) their model, a type of "creation model" which some have termed "progressive creationism." This seems odd since both accept cosmic evolution over giga-years and would appear to be quibbling over terminology. The question arises however whether millions of special creation (interventions) events might be indecipherable from theistic evolution empirically.[48]

---

[45] Obviously this assumes a high view of scripture (literal) but not an overly "literalist" or biblical fundamentalist" view (in which the 6 days of Genesis are interpreted as 6 literal 24 hours days).

[46] Also called cosmic evolution in which the universe evolves over a period of billions of years to its present state from an initial singularity. This would represent both special creation (of the original singularity including the selection and fixing of the constants of physics) and theistic evolution (the gradual unfolding of the material universe as demanded by the constants of physics including the effects of quantum irregularities and any chaos events which would be operative) as suggested in the book of psalms.

[47] E.g. First Life and Human beings at a minimum. The Cambrian "explosion" might be another that could be included and debated both from a scriptural and scientific perspective arguing to the best explanation or "best-fit" of the data.

[48] Dembski purports that special interventions in the form of ID are distinguishable in principle from the kinds of changes which could be produced by evolution even if it were theistically controlled or guided. This point remains unclear apparently even to Behe who has now raised it as a possible way in which to infuse design (through quantum irregularities in which God acts through secondary agent causation as a part of His Divine Providence).

Kenneth Miller and Michael Behe, based on Behe's recent chapter in *Debating Design* in which his view has changed more to Miller's view (except for continuing to embrace the concept of irreducible complexity in biology) apparently favor only the first (initial condition setting) and in this seem to apply their version of theistic evolution to standard neo-Darwinism.[49] Yet Behe holds that ID could have been imparted at various points in the history of life by "quantum events such as radioactive decay" or presumably other specifically undetectable physical processes, a position which appears to limit ID to secondary agent causation as part of Gods Divine providence, and one which to this author no longer comports with Dembski's formulation of intelligent design in science or Behe's previous view.[50]

Ross and Rana on the other hand are really advocating both special creation and a degree of theistic evolution, since the evolution of the cosmos proceeds according to initially chosen physical laws as well as various other special interventions (creation of the moon for example as a result of an asteroid impact with early earth) and then life is presumably introduced via at least one and then ostensibly multiple other episodes of supernatural intervention (special creation events) alongside or on top of natural evolutionary processes at the level of species. Miller, Jaki and other theistic evolutionists apparently posit only one creation event (initial fixing of physical constants coincident with the "Big Bang") from which the universe and life progressively develop (evolve) over time. Obviously, this event was "designed" so as to allow for all future development of the cosmos and life on earth, so-called "front-loading." Such a postulation however, is not

---

[49] See Behe's chapter 19, (352-370) in *Debating Design* and chapter 9 of Miller's *Finding Darwin's God*.

[50] Ibid, 357-358. As indicated above, Dembski holds that the "design" his theory identifies is not associated with law-like "front-loaded" processes which are the staple of "theistic" evolution. His "design" inference is to an unembodied designer who intermittently infuses new information into the already existent natural realm whether it be extra-universal and transcendent or from within the universe.

compatible with what Dembski and other ID theorists are advocating by their design theoretic (ID) formulation. They posit that the subsequent *design* was "infused" at various other points in the natural history of life on earth.

The Cosmic Factor

A critical and too infrequently appreciated concept in "Origins" is that life is inextricably linked to the development of the cosmos. Despite the fact that the history of the cosmos is not technically part of the "Origin and development of life" debate, it is indispensable to it in that multiple scientists have now demonstrated that life would be impossible were it not for a myriad of so-called Anthropic cosmic coincidences which make life possible on earth.[51] Therefore, any discussion of life's "Origins" is incomplete without taking into proper account the relevant implications which come from cosmology. This is one of the author's criticisms of the "Origins" debate thus far. It has too often occurred in a vacuum without proper attention being given to the demands of physics, chemistry and cosmology. Many Neo-Darwinists are insufficiently mindful of the severe limitations which cosmology places on their favorite theory. For example, it is now argued that the temperature and environment (i.e. heavy asteroid bombardment) of the early earth from roughly 4.5 bya to 3.9 bya was totally incapable of supporting any life even at the unicellular level. Persuasive data now exist which establish that biological activity by cellular organisms (without evidence of

---

[51] Guillermo Gonzalez and Jay W. Richards. *The Privileged Planet:* How Our Place in the Cosmos is Designed for Discovery. (Washington, D.C. Regnery, 2004); John D. Barrow and Frank J. Tippler. The Anthropic Cosmological Principle, (New York: Oxford University Press, 1988), (paperback edition). Brandon Carter, "Large Number Coincidences and the Anthropic Principle in Cosmology," *Proceedings of the International Astronomical Union Symposium, No. 63: Confrontation of Cosmological Theories with Observational Data*, ed. M. S. Longair (Dordrecht-Holland/Boston, U.S.A.: D. Reidel, 1974), 291-98.

biotic precursors) occurred at 3.5 bya and perhaps as early as 3.85 bya, leaving an incredibly narrow window of geological time for complex unicellular life[52] to arise from non-living chemicals as postulated by so-called "molecules to man" atheistic materialism. This *narrow window* of time is totally contrary to what would be expected from the standard neo-Darwinian model and needs to be adequately explained.

With the many cosmological limitations in mind and with the present inability to plausibly explain the natural appearance of "first life" on earth, it is logical to posit the special placement (insertion) of either new information and or matter in way of explanation.[53] The only reason not to do so and to test the hypothesis is if one has an a-priori objection which limits the possible explanation to only natural causes. The data at present does not support such a position and suggests that from the perspective of "Origins" science (if there is truly such a thing)[54] it would be fruitful to abandon methodological naturalism.

It is "cleaner" if one is arguing the case for special insertion

---

[52] Note that even the earliest unicellular life forms documented in the fossil record were extremely complex. See details in bibliography.

[53] This is more than Dembski (who concentrates on detecting design in science) for example is willing to posit with his theory of ID which is apparently limited to detecting the infusion of new information only, not necessarily new arrangements of pre-existing matter specifically, although this is the author's understanding of Dembski. See his *No Free Lunch*. **This author (on the basis of both scripture and the scientific findings to date) is positing that "first life" was specially inserted into the natural realm in the form of a complex unicellular entity which would represent both an incredible amount of new information but also a new "form" of already existent but rearranged matter (elemental C, N, O, H etc. into biopolymers and the first replicating unicellular organism) simultaneously.** Thus the necessary material, efficient and formal causes (causality in Aristotelian/Thomistic terms) for life on earth are provided by the first complex unicellular organism (matter and form) and the final cause (i.e. causality) is an immaterial transcendent first cause who establishes it by specifically intervening.

[54] Jaki for example would contest the very possibility of Origins "science" since the event(s) in question are not repeatable or measurable (quantifiable) in real-time. See his two works in the bibliography.

of either information (as ID might posit) and new rearrangements of matter (*forms*, as this author is advocating) to invoke a special creation event for "first life" in the form of the first unicellular organism and with it the *apparent* irreducible complexity of basic biopolymers DNA and RNA and the necessary cell membrane, then to invoke it later and frequently the way many ID theorists and some special creationists seem to do over gigayears of time and development in presumably multiple yet undisclosed or unspecified instances.[55] Note that ID critics; Miller, Ruse, Pennock and others have accurately asserted that ID advocates have not specifically stated when or how many times the "designer" intervened over the past 4.5 billion years while introducing irreducibly complex entities.[56]

From the perspective of creating an additional testable "Origins" model (one which includes specially inserted information and or entirely new entities) it would seem preferable to commit to a fixed number and kind of specially designed events in time and then to

---

[55] To date, ID theorists (unlike special creationists Ross and Rana) appear not to have detailed the number and type of presumed design insertions ("infusions" to use their terminology) they advocate other than Behe's three examples and recently Bradley and Meyer's additions of the Origin of Life and the Cambrian. Nor have they produced a testable "Origins" model like that of Ross and Rana. Dembski has resisted doing so apparently on principle. They continue to infer design at a biological level in patterns, what the author referred to as the "aggregate" earlier, and some specific circumstances (bacterial flagellum, clotting cascade, cilium and now apparently first life and the Cambrian but to date have otherwise avoided comments on dates and overall number or kind of design insertions, which makes testing the theory more difficult in this authors view. So far, they appear to have mounted a categorical argument in favor of design by use of Dembski's explanatory filter to select certain "objects" which are likely candidates for ID after surveying the field for patterns which display design. Some think this "knock out punch" strategy is an attempt to prove too much and prefer the side by side testing of two competing "models" arguing to the best explanation.

[56] See Pennock's chapter 7 (133) in *Debating Design* in reference to the exchange which took place between design theorists and Neo-Darwinists at the American Museum of Natural History, also accessible on-line. Note that Dembski argues that design is detectable in nature by studying "effects" which he has analyzed using the tools of information theory and statistical analysis. He makes no attempt to propose an overall "Origins" model.

compare and contrast the available and future data to see which model was more consistent with the data arguing to the "best fit", what might be called the one with the most "explanatory power" (as Sober, Meyer and Lipton have advocated in separate writings).[57] Presumably, one of the two would also be more predictive when looked at in retrospect (retrodictively). This is discussed below in more detail.

---

[57] Note that design theorist Dembski has opposed this proposal in principle in his book *No Free Lunch*.

6. An Origin's Model Proposal

The author's own **specific view** of "Origins" also seems to be changing more along the general theme of Ross and Rana's view (i.e. their RTB creation type model, but with only a few special creation events rather than thousands or even millions as they seem to infer)[58] with the possibility of some heretofore undiscovered natural physical/biological laws (ala Denton) also being possible.[59] This would provide a testable model with which to challenge a strictly neo-Darwinian framework and possibly also answer the dilemma of "first-life" which to date, neo-Darwinism has no answer for.[60] If however, a completely naturalistic explanation for "first life" is subsequently found, then Miller and Jaki's form of theistic evolution without subsequent special creation (intelligent design insertion) would appear to be vindicated. It is unlikely but not impossible if "first life" was naturally produced, that any other developments thereafter would have been associated with ID such as the Cambrian or the arrival of human beings. A naturally documented origin of first life would strongly suggest only one

---

[58] The Reasons to Believe Creation model (RTB) of H. Ross and F. Rana, partially available on line at their web site reasons.org and in their recent book; *Origins of Life*: Biblical and Evolutionary Models Face Off. (Colorado Springs, Colorado: NavPress, 2004).

[59] Which could represent a type of theistic evolution combined with at least "3" and maybe more instances of special creation (first life, the Cambrian "explosion" and Human beings at a minimum for example), in addition to microevolution in the form of natural selection acting on genetic mutation at the species level at the very least (which no one involved in the "Origins" debate seriously doubts; so-called microevolution that not even "young earth" creationists deny. The possibility of additional undiscovered natural/biological laws should not be dismissed as future research might delineate them and awareness of same could hasten those discoveries.

[60] Obviously, this could change with further research but at present seems unlikely. Note also that if only one scenario is involved, such as theistic evolution, it would be empirically impossible to differentiate between what God did (design), what chance did and what necessity (law) did. John Polkinghorne recently made this point in his chapter in *Debating Design*.

creative event coincident with the fixing of initial physical constants and the "Big Bang" It would not however eliminate design (as part of Creation) altogether since that question is clearly metaphysical. The design would be empirically undetectable however having been accomplished through the initial fixing of physical constants and through God's prior knowledge (omniscience) of what environmental and other chance effects would add. This would lie within God's Divine providence and be accomplished strictly through secondary agent causation. Such a circumstance however, would be very unpalatable to those orthodox Christian theists who take a high but not literalist view of scripture such as Ross, Rana and this author.

Notably, Ross and Rana (RTB model) accept microevolution and numerous episodes of special creation (in fact many more than the author is willing to), but they appear to be concerned with the approach of ID theorists who argue that design is identifiable by ruling out chance and or law in biology on a categorical basis through finding certain instances of specified complexity or "irreducible" complexity in patterns or in specific objects or "system by system". They are also somewhat critical of the ID community's notion of a non-descript "generic" designer which could be either extra-terrestrial (intra-universal such as ET) or extra-universal (God). In the words of Hugh Ross:

> "An ill-defined designer leaves a model so sketchy that it is difficult or impossible to test with evidence or predictions—a model too nebulous for scientists (and others) to trust. The reluctance of most I.D. theorists to openly discuss creation theology adds to scientists', politicians', and reporters' distrust. Many see the I.D. movement as a thinly veiled subterfuge for sneaking failed creation models back into public education curriculum." We at RTB want to encourage open discussion of various origins models. Additionally, we want to encourage greater, not lesser, critical thinking about origins. Neither political correctness nor blind faith leads to truth. A more

enlightened approach openly recognizes that nearly all of
science overlaps nearly all of religion. I encourage the pursuit
of any model that can withstand testing.[61]

   This author accepts the concept of specified complexity in
biology but thinks that the idea of "irreducible complexity" of
individual organisms, structures or systems in biology (after the
initial first life creation event of at least DNA, RNA and likely the
first cell) is unproven so far, and possibly an empirical "dead end."
Irreducible complexity may not be synonymous with biological
specified complexity despite the assertions of Behe and Dembski.
It is disconcerting that to date; the ID community appears not to
have made progress on their design theoretic research program[62] It
is certainly theoretically possible to falsify ID in any specific case
by finding a plausible but yet at present incompletely tested
pathway much as Miller and others have tried to do.[63] The reality
for ID theorists however, is that as presently articulated, neo-
Darwinism is virtually impossible to falsify scientifically since it
essentially requires proving a universal negative which is

---

[61] Hugh Ross, "More Than Intelligent Design" *Connections* 2001, volume 3, number 3.
[62] It is very possible however that ID theorists are correctly drawing attention to design
(specified complexity) overall in biology that is; in the aggregate even though it may or
may not be possible to do so at the individual organismal or systems level (object level)
using the concept of irreducible complexity as a stand-in for CSI (complex specified
information) and this should be further investigated. It is not clear yet to this author that
IC is a valid concept for CSI. If empirically demonstrated, individual instances of ID at
the organismal or systems level would provide a separate insight (a type of check if you
will) for what is already known on the basis of metaphysics, not a challenge to the need
for ultimate (final) causality which many philosophical naturalists misguidedly hope to
establish through science. It would also strongly suggest that the Creator/Designer
employed a tri-partite mechanism for development of life on earth; chance, law and
design. Christians would find this Trinitarian theme heartening of course.
[63] Note that Behe and Miller have debated back and forth on the cilium, the bacterial
flagellum and the clotting cascade both on-line and in the 2004 book *Debating Design*
and it is apparently still on-going based on their recent chapters in Debating Design.
Behe appears to be modifying his view more toward Miller however while still retaining
the concept of irreducible complexity in biology. See his chapter from *Debating Design*,
(357-358). It is therefore difficult to determine exactly what Behe now believes re:
"Origins".

effectively impossible, while ID as presently promulgated is theoretically capable of being refuted, a situation which is clearly unfair.[64] This much has been appropriately argued by Behe.[65] Thus the two are not on equal footing as testable hypotheses, a problem which would be rectified by comparing two well-delineated "Origins" models.[66] As presently articulated, biological ID (Irreducible Complexity) seems to side-step the very difficult field work, (prediction and well thought out empirical studies) that is required if we are to adequately compare various "Origins" models. This criticism has been frequently articulated by ID opponents perhaps justifiably.[67] Unlike type-ordinaire *operation science* which benefits from repeatability and reproducibility, "Origins" research as origins science is inherently inductive, and presumably non-reproducible and non-repeatable in time. Thus a different technique for comparison is in order; so-called explanatory power, or argument to the best explanation.

The author is impressed so far with Ross and Rana's RTB creation model[68] but would open it up to an additional possibility (biological law) which would lessen the number of times the Creator would specially create (ID insertion). In that regard, Miller's idea has some currency in arguing that it is even more impressive for God to have designed a "life-system" which would have some *built-in* capability to evolve/unfold (without the Creator/Designer needing to intervene specially for every modification), not only on the basis of natural selection but on the

---

[64] This in theory could be changed by comparing two different models however, each with its own predictions and then arguing to the best explanation.
[65] See his chapter in *Uncommon Dissent.*
[66] This is unlikely if Dembski continues to object in principle.
[67] Miller, Pennock, and others.
[68] Although it is exceedingly difficult to find the details outlined anywhere including on the Reasons to Believe web-site. The author attempted to contact both Ross and Rana by e-mail (unsuccessfully) for further details. They contend that recent discoveries in biology and astrophysics support their model over a neo-Darwinian one and thus further details of their model would be critical for comparison.

basis of designed laws although they need not be entirely undetectable as Kenneth Miller seems to require.[69] It would also seem to be more reasonable and consistent with cosmic laws to posit detectable biological law on top of natural selection and at least "3" Divine instances of special creation.[70] When combined with some degree of minor variation and natural selection the process would become tripartite in totality as follows;

1.) Special "Big Bang" creation event including selection of the initial physical constants, first life[71], and the geologically recent origin (30-100 thousand years) of human beings assuming roughly a 14 billion year old (byo) universe ± the Cambrian "explosion." (Note that if Theistic evolution alone is true, then one could dispense with all but the first).[72] This however, appears proscribed

---

[69] Kenneth R. Miller. *Finding Darwin's God*: A Scientist's Search for Common Ground Between God and Evolution. (New York: Harper Collins [Perennial Edition], 2002). Presumably he demands they be undetectable because he thinks neo-Darwinism is completely explanatory at present and sees no need or possibility that any subsequent empirically detectable laws will be discovered.

[70] The justification for any Divine Special Creative actions other than the original Creation event (Big Bang) is that a literal (but not literalist) reading of Sacred Scripture and for example Catholic Tradition appears to demand it. (A completely metaphorical reading of scripture and or a repudiation of Catholic Tradition would not of course which allows many Christian mono-theists to accept theistic evolution with only an initial Creation event at presumably the "Big Bang)." This is really Deism or Panentheism however which is incompatible with orthodox Catholicism.

[71] Yockey has estimated that the synthesis of even one average sized protein such as cyctochrome C by chance alone is as incredibly small as to be impossible, H. P. Yockey, "A Calculation of the Probability of Spontaneous Biogenesis by Information Theory." *Journal of Theoretical Biology* 67:377-398, 1978; Strait and Dewey (1996) found 1 chance in 10(75) and Bowie et al. (1990) found 1 chance in 10(63) for the spontaneous accidental formation of one functional protein both of which are virtually impossible in the geological time available. See full citation in bibliography. Also see Bradley's chapter in Debating Design (especially 336-337) including a discussion of Shannon information theory and its application to complex specified information (CSI) and the probability of accidental protein formation.

[72] As particle physicist and Anglican Priest John Polkinghorne advocates. See his chapter in *Debating Design* (246-260). He argues that if theistic evolution alone is true, it will be

by Sacred Scripture and Sacred Tradition (of the Roman Catholic Church e.g. Humani generis) which requires that the Human race be started by two literal first parents not a group of human beings.[73] It also appears unlikely based on the accumulated scientific data so far. Theistic evolution would also be eliminated by any non-Catholic Christian literal understanding of scripture, (any reasonably orthodox non-Catholic Christian rendering) and by a Biblical fundamentalist (literalist) reading which completely rejects macro-evolution (descent with modification) because of its belief in a literal 6 (24 hour) day Creation week.[74]

2.) Some theistic evolution in the form of cosmic and biological evolution according to known physical and possibly undiscovered biological laws the total extent of which is unknown at present particularly with respect to biological laws in which only natural selection is now identified.

3.) Some (call it quantum for lack of a better term) variation in the form of the mutation/natural selection mechanism of biology and the aggregate effects (at a cosmic level) of quantum uncertainties and chaos. This has the result of providing consistency between the cosmic and life "Origins" models which is reasonable if the universe has a transcendent unmoved mover or first cause of unlimited power and intelligence.

The real problem with "Origins" of course is that only a radical philosophical naturalist could: look at the "fine tuning" of the cosmos for life, the incredible complexity of biological life and yet conclude that it is not the result of a super intelligence in some

---

impossible to determine what chance, law or design have accomplished. See his discussion on 258 of the above.
[73] (Humani generis for example, #'s 36 &37), in order to avoid polygenism and the resultant inability to account for the original sin of Adam and Eve, a necessary and indispensable element of orthodox Christianity.
[74] A time period insufficiently long for any form of evolution to take place.

way or another[75] i.e., theistic evolution, progressive creation, or
some combination thereof.[76] That is what philosophical naturalists
or committed radical Darwinists expect people to believe and it is
simply incredible, unfounded and insufficient. It flies in the face of
everything we know about living life and it also contradicts first
principles. Radical philosophical naturalists are clearly vulnerable
in positing an explanation for the development of life which on its
face is insufficient to account for what we observe.[77] There is no
evidence that anything in real-time occurs without adequate cause
(causality in Aristotelian/Thomistic terms). There is similarly no
evidence that wide-sweeping morphological (phenotypic) changes

---

[75] The trajectory for "fine-tuning" of the universe is increasingly more not less "fine-
tuned" and the trajectory for the specified complexity of life continues to increase at an
astounding rate both in its variety and in its detail down to and including the smallest
level of existence (cellular, subcellular, molecular and it appears even atomic and perhaps
ultimately subatomic). The degree of specified complexity in the aggregate far exceeds
what is even really imaginable by one or any number of human beings working at it
continuously over their entire lifetimes. If philosophical naturalism were true, these
trajectories should be inverted since the incredible fine-tuning of the universe and the
staggering specified complexity of life would be the result of mindless processes i.e. the
result of unintelligence which somehow gave rise to a level of intelligence unknown in
human history. Such a possibility is simply logically and metaphysically absurd.
[76] Clearly this understanding is based on classical principles of philosophy beginning
with Aristotle and later expanded upon by Aquinas (and more recently by Dominican
trained physicist, engineer, theologian and Philosopher William A. Wallace. *The
Elements of Philosophy*: A Compendium for Philosophers and Theologians. (New York:
Alba House, 1977); & *The Modeling of Nature*: Philosophy of Science and Philosophy of
Nature in Synthesis. (Washington D.C.: Catholic University of America Press, 1996) in
which it was contended both by pagan and Christian philosophers that material reality
can not be its own first cause lest one engage in an infinite regression. See specific
references in # 2 for details. **Thus material reality must have an immaterial first cause
on the basis of philosophical principles.** This is consistent with the transcendent
omnipotent God of Christianity, Judaism and Islam.
[77] This could be easily rectified by their simply stipulating to the need for an immaterial
first cause of material reality in exchange for limiting science to methodological
naturalism and or by admitting en-masse that biological science (neo-Darwinism) says
nothing about the possible existence of a transcendent creator/designer and that
evolutionary biology is concerned only with the mechanism of species development on
earth over time. Then we could settle down to discovering which of the "3" basic
mechanisms were responsible (chance, necessity and or design). Under this rubric "self-
emergence" a possible fourth, would be subsumed under front-loaded design.

can be documented in real-time and thus their extrapolation to geological timescales in which giga year are involved seems logically unwarranted if not absurd in the absence of an immaterial first cause.[78] Real-time alone by our common experience (i.e. not geological time) is associated with nothing other than decay unless energy is expended to temporarily reverse/halt it according to the second law of thermodynamics.[79] In order for that to happen, something which is sufficient that is, truly capable of doing so, must account for the expenditure of energy that is required to temporarily halt or reverse the second law's effects at the organismal level. A transcendent (immaterial) first cause does so by providing adequate sufficient (efficient (causality) but philosophical naturalism does not since it lacks the necessary metaphysical reserves.

Metaphysically it is inescapable and beyond reasonable doubt that the universe and life are the products of intelligent agency if one has no a -priori bias against the existence of immaterial reality or in favor of philosophical naturalism.[80] This is completely consistent with everything we know about complex and specified entities in the aggregate.[81] Such things in

---

[78] If one assumes an adequate immaterial first cause, then such large-scale phenotypic changes would be acceptable under either special creation or theistic evolution. Yet, most radical Darwinists deny the existence of an unmoved mover or immaterial first cause for material reality.

[79] Modern astro-physics also predicts an eventual heat-death of the universe, in which all chemical/biological processes will cease.

[80] Particularly in light of the Anthropic "coincidences" now widely documented and the incredible increasing complexity which is being delineated at the molecular biological level.

[81] In every other setting, complex specified entities are in the aggregate easily discernable as designed and when applied to biology the concept dates to Orgel and Davies and then expanded on by Dembski. Perhaps it is simply impossible to determine design at the individual organismal, organ, or "systems" level utilizing the same technique. This remains to be demonstrated empirically. However, every object designed by humans is not completely composed of parts which are designed by humans. Consider a fireplace made partly of stone. Some of the parts are man-made and some are natural. An analogous situation (part specially created and part evolved) is not theoretically

our collective experience simply don't appear without adequate explanation which includes intelligent agency. Therefore, in the properly conceived "Origins" context, we are simply debating the details of how the universe and life might be explained or in other words how precisely an unimaginatively superior intellect accomplished it if we accept the overwhelming metaphysical and scientific evidence now available.[82] In the alternative, we are debating an absurdity of no consequence if everything we now know about life and the cosmos is the result of unintelligent random forces etc. It is violative of first principles to argue that intelligent agency (in this case human beings) can be caused by unintelligent forces or quantum events.[83] This is true even if we limit the discussion to Baconian (modern) notions of science which presumably includes efficient causation (not causality) but is questionably provided by neo-Darwinism alone. The burden of proof should be on those; who purport that the universe and life is not comparable to everything else we know with respect to adequate (sufficient) causality not those who claim that they (the universe and life) are comparable.[84]

In way of analogy: It's much like the difference between being given a new car (as a present) vs.: being taken to an

---

prohibited for the larger universe a situation which is compatible with an orthodox interpretation of Christian scripture as well (Psalms speaks of God "stretching out the heavens" which is consistent with the cosmic evolution that is part of "Big Bang" cosmology). In fact a combination of the two (special creation and evolution) might be demonstrative of greater artistic license.

[82] Many who do understand the implications are simply unwilling to accept the necessary alterations in their overall behavior and worldview which are implied by it.

[83] No effect can be greater than its cause by first principles.

[84] Once we regress back to first life in our inferential a-posteriori trail, a transcendent first cause must be postulated whether abiogenesis is possible or not, just as is the case for the existence of the initial singularity. This much is a matter of metaphysical necessity. The question of whether non-life can give rise to life through completely natural means without supernatural intervention is not one of metaphysical significance and says nothing about the possibility or necessity of an immaterial first cause for all of material reality including life, assuming that one has an adequate grasp of both science and philosophy.

automobile assembly plant and watch a car be systematically assembled part by part on an *assembly-line* over several days of time. At the end, you would have a new car. In the first case, one could determine that the car had been designed (clearly it is complex and specified as the ID theorists tell us) but you would know very little about what went into it, part by part. In the second case, you would also know it had been designed, but you would know a great deal more about how it was accomplished step by step and a great deal about the materials used. You would not however assume that the individual parts on the "conveyor belt" were being indiscriminately "morphed" one into the other but that someone else had provided the appropriate pre-provided "parts-list" for assembling a car from a pre-arranged and engineered design plan.

The car is a finished "form" or pattern existing first in the mind of the engineer which was then brought into reality by arranging other entities (the individual parts from the parts list) into the form of a car by reference to the design plan. The radical Darwinists (philosophical naturalists) fail to appreciate that they have confused the automobile assembly line and the parts list which includes the development of the idea and form of a car, and its design plan with their belief in a completely natural "morphing" of one entity into another without an agent (intelligent agency) capable of selecting in advance which recycled or original parts to utilize. This largely relates to their refusal to adhere to first principles of being and to their minimalist/reductionist understanding of the philosophy of nature.

To carry this analogy a bit further, it would be as if each part on the assembly line were capable of becoming (morphing into) any other part simply by trial and error rearrangement of its atoms and molecules with the proviso that in no case could any such metamorphosis interrupt the continuous movement of the line to a conclusion the line knew not what. The analogy to natural selection would be any potential rearrangement which would assist the assembly line in continuing to function where any inherently dysfunctional one would interfere with or stop the line (analogous to fatal mutations which are not "selected" due to their (favorable) reproductive advantage). Radical Darwinists would benefit from recognizing that in theory, the parts (organisms, organs, systems they study) could either be recycled or specially created, since by metaphysics, both require an ultimate explanation. The question for them is whether all the parts are rearrangements of prior materials or whether some have been specially provided. ID theorists contend that at least some have been specially provided. This in theory should be testable if true.

ID for its part, argues (e.g. bacterial flagellum, clotting cascade etc.) that some of the "parts" in the assembly line are manufactured from scratch and that some are recycled, which as noted above does not really answer the question that needs answering; where did all the parts come from in the first place? That is, ID infers that a designer was involved by positing that some of the parts were manufactured from scratch (directly designed) rather than being recycled (indirectly designed). They know that a "designed part" establishes a designer which can be likened to the God of Christianity. They should also know that even recycled parts have to be created (designed) originally lest one engage in an infinite regress which is incompatible with what we know from metaphysics and the philosophy of nature. Thus it can be argued that they attempt to prove the need for a Designer by the use of science (by redefining it to include goals, purpose etc. and thus opposing metaphysical naturalism) since it is clear to them that philosophical naturalists do not accept the need for a Designer (even of the universe), in principle. They have attempted to get an immaterial "first cause" back in the door by redefining science to include what Bacon called metaphysics in trying to defeat philosophical naturalism their real "enemy." Unfortunately they have provided an inherently inferior attack on philosophical naturalism (presumably their ultimate goal) and one which the author is in sympathy with. Tactically, one could argue that it would have been better to accept methodological naturalism for the purposes of doing science and then demonstrate 1) that neo-Darwinism cannot explain the appearance and development of all life on earth and 2) that any attempt to invoke formal and final causality as part of their scientific theories was to break the rules. Such flagrant breaching of the rules by neo-Darwinists and other philosophical naturalists is simply too widespread and common to detail.

Rather than Darwinists alleging that the Creationists and ID theorists were not "scientific" the ID theorists could have legitimately accused the Darwinists of not practicing proper science and contaminating it with religion and philosophy. Then, the philosophical naturalist neo-Darwinian scientists would justifiably *stand accused of "mission creep" over into metaphysics*, and guilty of violating their own rules. ID theorists apparently were unsuccessful in waging that type of argument and instead chose another course. They are still attempting to reinsert design into science without perhaps adequately indicting their competition for breaking the rules of methodological naturalism. Given the control that the Darwinian establishment has on science however, one cannot be too critical of ID theorists or their tactics.

While biological ID if true, would presumably establish an extra-terrestrial designer, it need not establish an extra-universal one which would be necessary in order to explain the existence of the parts. Any extra-terrestrial-intra-universal origin of the "parts" (even ET derived) would need an extra-universal first cause since ET and the parts need to be material.[85] This much is true on the basis of a study of the philosophy of nature and metaphysics and should simply be stipulated to by both Darwinists and Intelligent

---

[85] Otherwise it can be argued that one is resorting to spiritual (supernatural) assistance.

Design Theorists alike.[86] If biological ID is false[87], the possibility of an extra-universal designer is not eliminated since an extra-universal "law and or chance driven" design could have been selected by an extra-universal transcendent being with supernatural intelligence including the possibility that heretofore undiscovered laws (in addition to natural selection) were employed which may or may not ever be empirically detectable.[88] What all of the aforementioned ultimately reduces to is the following question: **can material reality be its own first cause?** As indicated above this is ultimately and primarily a metaphysical (philosophical) not strictly scientific question and one which metaphysics demands be answered in the negative.[89] It also points to one of the fatal weaknesses of neo-Darwinism as a complete and adequate cause of the "Origin" and development of life on earth.

---

[86] In the absence of such a stipulation (agreement) [which seems extremely unlikely given the large number of philosophical naturalists among the biological community], both should agree that ultimate (final causes) lie outside the realm of science and that all such metaphysical/theological claims will be avoided when conducting research or commenting on the purely scientific realm. This would require re-writing and re-editing virtually all of the works of modern neo-Darwinians which they would likely reject. ID theorists would then be completely justified in re-introducing design into science as one of possibly 3 or 4 ways in which the development of life on earth might or might not be accounted for. Similarly, self-organization theories such as Kaufman proposes would be epistemically equivalent to chance, necessity and design, but would in no way affect the metaphysical question of final causality.

[87] Where ID is understood as referring to specific "insertions" of either information and or matter into the natural realm either by a Supreme Being or some other presumably extra-terrestrial but intra-universal source.

[88] On the other hand, it might be possible to detect them along the lines of Denton's recent articulated view in which he invokes biological laws which are said to apply to the quaternary folding of proteins. This is clearly an area for more research. See Denton's chapter in *Uncommon Dissent*.

[89] Although the data from contemporary astrophysics are completely compatible with the metaphysical assessment in that the "Big Bang" suggests a finite universe in space and time which originated in a singularity. This is most compatible with a transcendent immaterial first cause as called for in Christian monotheism but also can be left a "mystery" for those favoring philosophical naturalism and presumably atheism/agnosticism or those unwilling to commit to the logical implications and alterations of behavior.

Simply put; until the dilemma of "first life" can be solved that is, until it can be demonstrated that life (organic chemistry) can arise naturally from non-living (inorganic chemistry) chemicals, naturalistic explanations for "Origins" simply engage in a material regression which ends at "first-life." Natural selection itself requires "first-life" to exist, a reproduction/replication system from which to reproduce and in so doing ultimately undergo genetic mutation. Only after mutation provides a survival advantage can natural selection be said to have "selected" anything. At that point it can be said to offer a survival/reproductive advantage and be "retained" through natural selection.[90] **Without an explanation for _first life_, we have no explanation for why organisms compete for resources and for the most suitable mates or why they "struggle" to survive at all.**[91] They might just as well all gradually undergo extinction which it appears has been happening progressively and slowly since the Cambrian. Perhaps more importantly, even if we assume that "first life" can eventually be demonstrated to derive from simple inorganic chemical processes, it doesn't eliminate the possibility that life on Earth is a result of the initial setting of the constants of physics, a conclusion which would then be inevitable and one which is still compatible with what is already knowable on the basis of metaphysics.[92] This would certainly be compatible with many versions of theistic

---

[90] Implicit in the neo-Darwinian synthesis is that there can be no goal in mind and no way to "plan ahead." Even the word "retained" carries the subtle implication of purpose, of design.

[91] This is a point which is not adequately appreciated or addressed by Neo-Darwinists who simply assume that organisms will compete for resources and that those which superior characteristics will have a reproductive and thus survival advantage (so-called natural selection). They never address the foundational philosophical question of why living things would compete at all. This proposition assumes that some higher law is at work which demands that living organisms will compete for resources. The radical Darwinist fails to provide a sufficiently powerful (in the ontological sense) explanation.

[92] According to Kenneth Miller a Darwinist Christian, this scenario places God in the most exulted position, an argument which has great merit and attractiveness in isolation of Sacred Scripture. Special Creation of "first life" is also compatible with the data from metaphysics.

evolution although it is problematic for those who hold to an orthodox view of Sacred Scripture and the Sacred Tradition of the Catholic Church.[93] Were it not for that however, the theistic evolutionary view of Miller and Jaki would seem to be favored if and when the dilemma of "first life" is ever solved <u>naturally</u>.[94]

The author proposes that two "Origins" models be compared including their ability to account for the historical record as it is presently understood in light of the relevant data from astrophysics (cosmology) the geological column, fossil and molecular biology (DNA sequence and eventually amino acid and three dimensional detailed protein analysis as is underway under the rubric of proteinomics) as well as their respective abilities to make predictions which can guide future research and discovery. Such a comparison should make one of the two the clear "inference to the best explanation" that is, the one with the most explanatory power as articulated by Sober and others.[95]

---

[93] Where an orthodox Christian literal exegesis of the Old & New Testaments at a minimum is required $\pm$ Sacred Tradition of the Catholic Church (e.g. Humani generis #'s 32, 34, 35, especially 36 & 37) but not a fundamentalist biblical literalist one which rejects theistic evolution in principle. If first life did not require special Divine intervention, then so-called theistic evolution is a virtual certainty since then non-theistic evolution alone is impossible on metaphysical grounds.

[94] John Polkinghorne could be included along with Miller and Jaki. The author's present operating assumption however is that it will remain a mystery thus keeping the possibility of special creation (ID insertion) of "first life" a viable one. It should be actively investigated by empirical means however along with a search for natural explanations of the Cambrian (in contrast to Meyer who invokes special creation [design] for the Cambrian) and Human beings (in contrast to Ross and Rana who postulate special creation of human beings [design]) at a minimum. Each of these is inherently more testable (comparable for inference to the best explanation) than is the case for irreducible complexity of specific structures or biological systems as advocated by Dembski and Behe to date where proposed pathways are difficult to test empirically without affecting the results due to investigator interference and where real-time constraints offer a theoretical reprieve for naturalistic explanations.

[95] P. Lipton. *Inference to the Best Explanation*. (New York: Routledge, 1991).32-38; Elliot Sober. *The Philosophy of Biology*, 2nd. Ed. (San Francisco: Westview Press, 2000), 44; Stephen C. Meyer. "The Cambrian Information Explosion" in *Debating Design*, 386. This technique is most appropriate for solving dilemmas which are inherently a-posteriori

For starters, the tripartite model in which "first life", the Cambrian, and Human beings are assumed to be the result of special creative ID acts (occurring on a backdrop of neo-Darwinian evolution) should be proposed and investigated as a competitor for the standard neo-Darwinian paradigm alone. A high view of scripture and the present data from biology and astrophysics makes this a logical alternative to the standard evolutionary model and one which should make unique and markedly different predictions and retrodictions by which further empirical research can be compared.[96] Obviously, in light of what is already known on the basis of metaphysics and revealed theology, if special creation of none of the most likely candidates can be inferred by a consideration of the data present and future (or additional special interventions which others might propose[97]), it would be reasonable to assume that theistic evolution (rather than special creation) was the mechanism selected by the extra-universal

---

(inductive) in nature and which at the most can generate only high degrees of probability as opposed to cognitive certainty. The "Origins" debate is clearly one of these.

[96] For example, the two make completely different predictions about the possibility for abiogenesis by natural means, about the possibility that the major body plans could have arisen through only natural means in the time allotted and with respect to whether human beings were specially created or derived physically and directly from a bipedal primate ancestor. Some preliminary work has been done which is applicable to all three areas and can be found in the bibliography.

[97] Such as the much more extensive RTB creation model of Hugh Ross and Fazale Rana admittedly much more difficult to test. In developing his testable Creation Model, the author however has selected the three most likely specially created (ID interventions/infusions) events in the history of life on earth, both from the perspective of Sacred Scripture and the Sacred Tradition of the Catholic Church as well as the data from biology as presently understood. This model does not depend upon a literalist reading of Genesis or any other related scripture, however it does depend upon a so-called "high view" of scripture and repudiates an entirely metaphorical rendering of the relevant passages such as would be presumably needed in order to accept Polkinghorne or Miller's view. The author assumes for example that Genesis provides actual history in some real sense, albeit written in the language and style of the period as articulated by Catholic magisterial teaching, not simply a metaphorical or mythical parable designed to make a conceptual point relevant only to salvation history.

(transcendent) intelligent agent responsible for the universe.[98] Under such a scenario, neo-Darwinism (of the non-radical form described by Miller) would appear vindicated as the empirically undetectable way in which the Creator effected/produced the entire created realm and yet allowed it to appear undesigned. **The only question left to ask would be why the Creator chose to make the creative activity empirically undetectable.**[99]

---

[98] Some would disagree however, e.g. Dembski who has argued that the Designer's likely behavior is unpredictable. See his *No Free Lunch*, 362 in which he argues that ID should not be subjected to the prediction of specific instances of design since the behavior of an innovator/inventor/designer is inherently unpredictable and unknowable. The author postulates that the designer is not simply a mechanic and innovator but a consummate artist who intervenes at various stages of natural history as part of the creative artistry of truly creating out of nothing, in addition to the rearranging that such a designer would share with human artists. The master designer would be capable of a degree of artistic expression unimaginable to humans both in the selection of the physical constants "front-loaded" into the universe and the special artistic interventions employed along the way.

[99] One possible answer is that the need for a Creator/Designer is evident from a study of the philosophy of nature and a proper understanding of metaphysics and Divine Revelation. Another possible answer might be in order to be certain that faith was not based primarily on empirically derived sense data alone but remained a gift through grace. This would accent the role of grace and would make both the book of nature and the book of scripture important means by which to approach the Creator. Design would thus be appreciable in the aggregate for all to see and appreciate but not individually, leaving something for scientists to ponder in detail. Some biblical scholars and others including the author contend that multiple scriptures are incompatible with this view while others do not, e.g. Rom. 1: 20, Psalm 19: 1 and many others, a subject which exceeds the scope of this work.

50

7. Conclusion

Despite the aforementioned, the author is convinced that theistic evolution as thus described[100] will never be vindicated in the scientific realm. A careful analysis demonstrates it cannot be supported in the philosophical or theological realms either. The writer's position is arrived at primarily from a consideration of the scientific evidence, both physical and biological but also as a result of a detailed analysis of the philosophical and theologically relevant supporting data. Both of the latter on careful examination are virtually incompatible with theistic evolution as promulgated by Miller and others. Theistic evolution to date has no adequate biological explanation for the arrival of the first human male (Adam) via physical generation from sub-human bi-pedal primates.[101] Perhaps more importantly, it has no explanation at all for the presence of the first human female generated in some tangible/physical way from Adam, an absolutely necessary requirement of orthodox Christianity and orthodox Catholicism as consistently and continuously promulgated by the Magisterium of the Catholic Church.[102] Since the actual physical existence of an initial set of first parents is irreducibly linked to one of the "3" core

---

[100] As advocated by Miller and Jaki.

[101] With every passing day, it appears on the basis of molecular biology including Y chromosome and mitochondrial DNA studies that there exist no pre-human bi-pedal primates in the fossil record which are directly related to humans. While it is conceivable that some future "missing link" will be found, the trajectory of recent discoveries makes this unlikely. The bizarre problem of accounting for how Adam could be the natural child of bi-pedal primate parents taxes the imagination and must ultimately be regarded as patent nonsense.

[102] See Humani generis of Pope Pius XII for relevant paragraphs which address the need to account for 2 actual first parents, the rejection of polygenism and the need for Eve to be generated by a supernatural act of God from the body of Adam in some physical way. Only those exegetical methods which employ a low view of scripture and reject the constant teaching of the Catholic Church and all of orthodox Christianity can reconcile the completely natural generation of human beings from non-human bi-pedal primates. This is perhaps the most significant defect in Miller's theory with respect to his attempt to make it compatible with Catholic Magisterial teaching, which it clearly is not.

principles of Christianity (the fall from grace), theistic evolution appears eternally and irrevocably implausible. If one disregards the Sacred Tradition of the Catholic Church including constant Magisterial teaching as well as the high view of scripture which is an integral part of orthodox Christianity, then theistic evolution remains a possibility but one which is purchased at too high a price. It requires destroying at least one of the foundational (core) principles of orthodox Christianity, the "fall from grace" of an initial set of first parents to whom the "fall" can be attributed. What remains is a "hollowed out" Christianity in no need of a redeemer and thus no place for the actual physically crucified Christ. Such *Christianity* exists in name only.

With respect to "Origins" we are thus left at the very least with some form of special creation of the universe, first life, and human beings if we are to adequately account for the data from modern astrophysics, biology, orthodox Christianity and the philosophy of nature. To argue otherwise is to be committed to an irrational a-priori bias totally in favor of philosophical naturalism despite good evidence to the contrary.[103] The best fit of the data at present is that at least "3" separate instances of divine special creation have occurred in the history of the cosmos, and likely many more. This conclusion is justified by employing what the philosopher of science Sober has called for, despite the fact that to date in his estimation; no such case had been successfully made.[104] This analysis represents a beginning.

---

[103] Including the best evidence from science, philosophy and theology.

[104] See Sober's essay in; Dembski, William A. and Michael Ruse. *Debating Design*: From Darwin to DNA.
Cambridge: Cambridge University Press, 2004.

8. Appendix Summary of Major "Origins" Issues

1.) It is clear that "Origins" issues involve at least "3" fundamental or foundational disciplines; science, philosophy and theology. While contemporary science is usually seen as the major source from which to draw relevant data and conclusions in "Origins", it is a mistake to think that science alone can adequately address the issue. This is true whether considering the origin of the universe or life itself. Both topics require input from philosophy and theology in order to adequately account for the diverse nature of the related subject matter. For most of the past 400 years since the time of Francis Bacon, and particularly in the past 146 years since Darwin's publishing of; *On Origin of the Species*, (1859) science has predominated the discussion of "Origins" by rigidly enforcing *methodological naturalism* and by attempting to weaken the role of philosophy as a foundational discipline. In so doing, "Origins" debates were re-cast as science (facts) vs.: religion (opinions) to the exclusion of philosophy where science alone was the obvious winner. As a result of the post-Enlightenment growth of philosophical naturalism and the acceptance of methodological naturalism in modern science, any discussion of goals, purpose, meaning, and ultimate design in either the universe or life itself were eliminated, in principle. Yet, many scientists continued to "insert" their "new-found" materialistic *worldview* assumptions back into their favorite scientific theories and commentaries thus creating a full-blown materialist metaphysics (philosophical naturalism). The "Origins" debate is foundationally a philosophical (metaphysical) one primarily, which utilizes data from science and theology to "fill-out" details, the basic pattern of which can be demonstrated by an understanding of the philosophy of nature and metaphysics at the first and second degrees of abstraction.

2.) Bacon, Hume and others effectively removed Aristotle's formal and final causality from science and arguably efficient causality as well. This left science with a sterile method which primarily considered only material "causation" and *physicalist* explanations irrespective of where the trail tended to lead a concept and approach which in itself is "unscientific." It was particularly effective from the perspective of standard "operation" science in which explanations were sought for "how things work" and where measurement to increasing levels of accuracy were sought, particularly at the astronomical, molecular and atomic levels. It also markedly increased overall ability to make use of technological applications grounded in basic science. Unfortunately, it also resulted in overly materialistic (often mechanistic, and eventually magical) thinking[105] which tended to "spill over" into a full blown acceptance of materialism as a *worldview* that is, in the acceptance of philosophical naturalism where reality is limited to the material realm only. Thus, many scientific naturalists (who should have limited their theorizing to methodological naturalism only by their own rules of proper conduct) began to extend their view of science beyond its natural boundaries effectively creating full-blown metaphysical and in some cases theological explanations in the intentionally created intellectual vacuum of true metaphysics and theology.

It will be impossible to make progress in "Origins" until the proper role of philosophy is again appreciated, particularly an Aristotelian/Thomistic view of the philosophy of nature (updated in light of modern scientific findings)[106], and a reacceptance of the role of metaphysics which since the time of Immanuel Kant has languished. The death of classical metaphysics can largely be dated

---

[105] Where it is assumed that all material effects have only material causes despite ample philosophical evidence to the contrary and where the first principles of being are essentially rejected.

[106] William A. Wallace. *The Modeling of Nature*: Philosophy of Science and Philosophy of Nature in Synthesis. (Washington D.C.: Catholic University of America Press), 1996.

to Descartes' "turn to the subject" despite the fact that he no doubt did not intend as much. This was continued by others in various ways and degrees particularly Emmanuel Kant's mistaken cosmology and repudiation of metaphysics and Hume's promulgation of what became known as British (logical) empiricism and skepticism which appeared in Bacon and Darwin and later more widely embraced by various other logical positivists in the United States and elsewhere. All of these had in common a basic assumption of materialism as an over-arching *worldview* and a denial of immaterial reality including God, and life after death.

Thus the limitation of science to material *causation* (methodological naturalism) only (as if it could be completely explanatory, rather than the classical understanding of material, efficient, final and formal causality), resulted in an almost obligatory assumption of philosophical naturalism. Coincident with it was the necessary overall skepticism of Hume and those who followed. Thus the fundamental mistake made in "Origins" research since the Enlightenment was one of selecting too limited a philosophical or foundational approach, one which created unacceptable a-priori bias in favor of only materialistic explanations.

3. As a consequence of # 2 above, modern neo-Darwinism is the contemporary and accepted form of materialism in Biology (that is a larger metaphysical theory or construct not simply a theory of mechanism), in which life is explained by invoking only material causes. It purports to explain the origin of all life on earth and its complete development by entirely natural means over eons of time by invoking a progressive but "undirected" and *unguided* evolutionary process (so-called "unintelligent" evolution) based in genetic mutation which is then acted on by natural selection. This is a form of inheritance based evolution which is modified by various chance processes (environmental and organic including point mutations, lateral gene transfer, copying errors, deletions etc.). <u>The process is primarily one in which "mistakes" are utilized by organisms to their practical advantage</u>!

Those alterations which confer a survival advantage are then passed to subsequent generations. According to neo-Darwinism there is no room for purpose, design or goals in the exclusively naturalistic evolutionary (unintelligent) process. This process is said to involve only chance and necessity (law, in the form of natural selection), but not "design" or purpose, despite the fact that no explanation is given for the "law" part of natural selection. Every law requires a "law-giver" and this is also true of natural selection.[107] **Such a "law" begs the question of why organisms compete for suitable mates and to "stay alive" so as to pass their genetic material on to the next generation.** They might just as well simply all go extinct as continue in existence, if Darwinian naturalism is true. The answer to this question no doubts lies in the explanation of "first life" which to date has no adequate naturalistic explanation and likely never will. In this sense "natural selection" is not really a "Law" in the usual physical sense but a superficial description in which the necessary or "sufficient" cause remains unknown. Had efficient, formal and final causality not been eliminated by modern science (methodological naturalism), this inadequacy would be readily apparent. That is to say the "law" portion of *natural selection* acting on random genetic change is simply not adequate (sufficient) in order to explain the entire development of life on earth. This should be obvious on the basis of philosophy and a reconsideration of the "4" causes of Aristotle/Aquinas. Put differently, the mutation/selection mechanism does not provide efficient causality and it ends in

---

[107] Even if mutation and natural selection alone could explain the development of all life on earth, the fact that it is based on taking advantage of naturally caused "mistakes" strongly implies design when one looks at the incredible panoply of diverse life on earth. "Mistakes" in real-time are usually negative in their effects however.

making all phenotypic (organismal forms) expressions mere illusions (appearances) rather than actual physical realities.

Ultimately, the problem is that radical Darwinian science (also referred to as philosophical naturalism or [PN]) has lost its foundation/anchor in physical reality. One wonders how "life" can be completely explained by a process which is totally dependent on chance including "mistakes" which are then "utilized" (selected) to the organism's advantage, if the mechanism is actually "undesigned" that is unintended by any sentient agent. Moreover, how can anything truly be selected in the absence of sentience if none is provided for by the theory without engaging in language deconstruction and verbal absurdity? Perhaps more befuddling, in no other situation do we invoke "mistakes" when attempting to explain complex and specified patterns. It is much more logical to hold that the complex specified patterns we observe in nature are actually rather than apparently designed especially if it can be demonstrated that some are the result of utilizing "mistakes" for gain. Only if one assumes philosophical naturalism would this not be the case.

More recently (roughly 50 years) discovered data in biology make exclusively naturalistic explanations (first life and subsequent development) extremely suspect due to the incredibly complicated (so-called specified complexity) nature of biological organisms and related "systems" and subsystems [including biochemical pathways], down to and including the molecular and atomic levels where there exist actual and literal "molecular machines" many degrees of magnitude more complex than anything ever produced by humans. It is now apparent that biology in general is literally packed with so-called "specified complexity" (CSI) which with every passing day becomes more not less evident. The trajectory is clearly toward increasingly greater and greater degrees of specified complexity the origin of which is in question just as is the increasingly apparent incredible degree of cosmic "fine-tuning" of the universal constants of physics needed to support life itself.

The growing dilemma with regard to specified complexity is the problem of explaining incredibly unlikely (improbable) and apparently sudden (in geological time) insertions/infusions of information into the natural realm and the tracing of the inferential trails that this information has made, as Dembski has rightly argued. In theory it is possible that the origin of such information will be found to be traceable to an extra-organismal and abiotic source which would be impossible assuming a paradigm of methodological naturalism. This work is on-going at present primarily by ID theorists and the few others who have proposed alternate "Origins" models to rival the standard neo-Darwinian one. However, it is also possible that we will never uncover the origin of specified complexity in biology, a point Dembski makes in his final chapter of *No Free Lunch*.

Nevertheless, ID theorists seem to have found a way of formalizing what is already inferable from the philosophy of nature and metaphysics; that highly complex and specified arrangements of matter are never due to chance processes and or those that involve fundamental natural laws both of which are centered in the nature of matter itself. The "design" that ID proponents are advocating is inherently unembodied (immaterial) and extra-organismal. The question remains whether they have established design at the aggregate level in biology only, or whether their design theoretic framework can document design at the individual, object or systems level in addition.

4.) Contemporary cosmology now suggests a finite universe in which both space and time were initiated (created), beginning in a singularity of virtually infinite density roughly 15 billion years ago as scientists calculate time. At that point, all matter/energy/space/time were "condensed" into an almost

infinitesimally small point from which the universe expanded in
size and evolved (enlarged and became diversified) to eventually
form stars, galaxies, planets etc. In the first "3" minutes after the
"Big Bang" singularity, virtually all the mass/energy existed in the
form of hydrogen (the lightest known element), a small amount of
helium and some deuterium. Later with the birth and death of first
and second order stars, the heavy elements critically necessary for
all living things were produced including iron the all important
element found in hemoglobin.[108] This has critical implications for
the "Origins" of life controversy in that the realities and limitations
presented by cosmology must be taken into account in any and
every "Origins" model proposed.

The general theory of relativity (GR) has now been tested
empirically and found to accurately describe the physics of the
universe to greater than 14 places of the decimal (less than 1
chance in 100 trillion that it is mistaken) and the inflationary Big
Bang which flows out of GR and predicts an expanding universe is
likewise established beyond reasonable doubt on the basis of
astrophysics. Unfortunately, the "origins of life" debate including
the complete development of life on earth, historically has not
adequately considered all the relevant implications of modern
cosmology. When proposing "Origins" theories, the origin of the
universe, its cosmic development over time and the implications
for first life on earth are critical to a proper consideration of
biology particularly evolutionary biology.

**5.) The philosophy of nature and metaphysics adequately
establishes that the material realm (universe) must have an
immaterial first cause**. It should be obvious that one is forced to
accept an actual infinite material regression if materialism
(philosophical naturalism) is true. It has been demonstrated beyond

---

[108] The hemoglobin molecule for example is responsible for most of the oxygen carrying
capacity in the blood of humans.

reasonable doubt that one cannot cross an actual infinite since one could never arrive at the present.[109] Accordingly it is also apparent from a detailed study of the philosophy of nature that the material realm cannot be its own first cause. Thus, it is unnecessary and unwise for science to be placed in the position of proving the existence of the first unmoved mover or the first cause. Having said that, a modern Teleological argument in light of contemporary science is certainly supportive along with the other classical and updated arguments for the existence of God. As an aside, the modern Kalam cosmological (philosophical) argument as articulated by Craig is a persuasive way in which to utilize the data from modern science in what is ultimately a deductive philosophical proof for the existence of God.[110] Briefly it is as follows:

1. **What begins to exist must have a cause.**
2. **The Universe began to exist.**
3. **Therefore, the Universe has a cause.**

In the past, some atheist/agnostic philosophers of great renown e.g. Anthony Flew have found the various arguments (cosmological, ontological, teleological, moral etc) for the existence of God to be unconvincing. Flew however has recently indicated that he is now open to a creator God on the basis of a reconsideration of the Design argument particularly in light of the "fine-tuning" of the universal constants of physics and the incredible degree of specified complexity which is found in biology. This is no small matter, since Flew has been the most outspoken atheist philosopher in the past and one to whom most

---

[109] See the writings of William Lane Craig and J.P. Moreland on this subject, nicely outlined in their respective works detailed in the bibliography.
[110] The Ontological argument of Anselm as modified by Plantinga's "all possible world's" thesis is also persuasive as is the Moral argument, (as articulated by C. S. Lewis and St. Thomas Aquinas in the 4th of his 5 ways.

proponents of philosophical naturalism have appealed for support. Dawkins would be an example of a radical Darwinist who is also an atheist and who previously derived intellectual encouragement from the arguments of Flew.

6. Contemporary Intelligent Design (ID) theorists have attempted to reintroduce "design" into biological science as a third type of mechanism (in addition to chance and necessity [law]) in way of explaining the origin and development of life on earth. This represents a redefinition of science (from its 400 year old tradition of methodological naturalism started by Francis Bacon to a more Aristotelian view in which all "3" mechanisms (chance, law and design) are considered part of science). It also reintroduces the possibility that efficient, formal and final causality (which Bacon rejected as metaphysics) will potentially again be considered part of science. These attempts have been completely rejected by radical Darwinists and many other physical scientists, in principle. They argue that science is synonymous with methodological naturalism, a position which is reflective of only 400 years of human history. Prior to that time, it was not the case and presumably need not be now other than by a-priori commitment to the status quo steeped as it is in materialistic philosophy (philosophical naturalism).

ID (despite advocating the reconsideration of design as part of science), is not an attempt at constructing a competing "Origins" model with which to compare and constrast the standard neo-Darwinian one, unlike the extensive RTB creation model for example. Detailed examples and models to date have not been part of the ID theorist's work or stated interest. They are primarily concerned with establishing the evidence for design in nature (through science) by documenting its effects through the utilization of statistical and information theoretic techniques. Their marker for design is *specified complexity* which they argue is never demonstrated in the absence of intelligence and by inference and ultimately an intelligent agent, ostensibly an immaterial (unembodied) one. To date this attempted recognition of design has been primarily at the aggregate or "pattern" level and a few specific objects and or systems such as the bacterial flagellum, the cilium, and the clotting cascade in which the presumed specified complexity is said to exhibit irreducible complexity [IC] (a biological type of specified complexity according to ID theorists). Whether the concept of IC is a valid "stand-in" for or type of specified complexity remains to be seen and at present is being hotly debated in biological circles. Many biological scientists are willing to grant that biological systems, organisms and objects "appear" to demonstrate specified complexity but not necessarily irreducible complexity and not necessarily <u>actual</u> specified complexity. So far it appears that proving IC is a matter of refuting proposed naturalistic pathways rather than providing positive evidence that specific infusions of ID have occurred at specifically identifiable times and places in the form of particular new organisms or systems.[111]

Some scientists e.g. Dawkins allege that any actual design which this *specified complexity* demonstrates is only apparent not real, an argument which seems not only counter-intuitive but inadequate on its face. **The responsibility for demonstrating that what is obvious is only apparent and not real or actual should reside with those who posit that we are being deluded by our senses and our integration of sense data in believing that biology is designed** (exhibiting and possessing specified complexity). Dawkins's formulation is heavily dependent on the logical empiricist school of Hume and Kant's idealism including

---

[111] Dembski's method of "proving" design in biological systems is dependent upon an approach which utilizes information theory and mathematics to categorically eliminate chance and necessity and thus by process of elimination identifies design as the only option "left standing." It is inherently deductive and empirical only in the sense that the factors which are selected for input into Dembski's explanatory filter are purportedly based on actual biological data such as the number of known proteins which constitute the bacterial flagellum.

his assertion that sense data is inherently untrustworthy. Science as practiced in modern times is much more Aristotelian/Thomistic where sense derived data is the accepted and valid way in which to gain empirical knowledge of the material world. In other words, the latter position is much closer to the realist position that actual (practical) scientific investigation utilizes. The Humean/Kantian view makes all scientific activity ultimately ungrounded in reality which is obviously unsatisfactory for any realist scientific enterprise. The Humean tendency also results in many scientific hypotheses bearing little if any resemblance to reality such as multiples universes or imaginary time in cosmology. People such as Richard Dawkins bring tremendous philosophical (foundational) assumptions to his version of science which pre-determines the possible outcomes. He assumes both methodological and metaphysical naturalism (materialism) and thus only materialistic explanations are possible under his rubric. Too often he speaks as a radical philosophical naturalist rather than a biological scientist who employs methodological naturalism in "doing science."

   7.) Scripture plainly indicates that the witness of nature in all of its glory is an obvious demonstration of the existence of God. For example, St. Paul in Romans 1: 20 writes as follows: *"Since the creation of the world His invisible attributes, His eternal power and divine nature, have been clearly seen, being understood, through what has been made, so that they [men] are without excuse."* The Old Testament also referred to the evidence that nature provides for establishing the existence of God. The Psalmist in Psalm 19: 1 wrote: *"The heavens are telling of the glory of God; and their expanse is declaring the work of his hands."* It is therefore reasonable (if scripture is a legitimate attempt by God to communicate with his creatures)[112] to look for

---

[112] An assertion which is born out by a careful study of the actual historical, prophetic and factual claims made in scripture and which have been found to be verifiable through the use of modern dating techniques, archeology, and other modern techniques of manuscript, papyrus and textual analysis.

evidence of God's creative work in the natural realm. Since God created man in His image, it is also reasonable to look for those aspects that man shares with God in order to direct the search for evidence. The 2 obvious characteristics which man alone shares with God are his intellect and will what is commonly referred to as soul. God has an immeasurable amount of intellect (omniscience) and a perfect will. It is therefore reasonable to look for evidence in nature of His intellect at work through His perfect Divine will. This leads to the suspicion that the universe and life will be extremely intelligently made which is equivalent to what is found in nature through modern scientific investigation, an extremely complex and specified biological realm and an intricately arranged "finely tuned" universe which is "just right" for life, something which is predictable of an unimaginably willful agent.

The data from Divine Revelation then must not be neglected in formulating an overall "Origins" model as is the case with modern cosmology (which has been demonstrated to be indispensable to the explanation of life on earth). Despite the fact that many scholars have held that scripture particularly Genesis, Exodus and other related passages are not primarily intended to be science texts (a point which is undoubtedly true), both make "fact claims" which are eminently testable empirically such as the material realm being created ex-Nihilo by a transcendent immaterial entity and that human beings are the very privileged end result of a process which had them in mind from the beginning. Whatever actual historical information is contained in Sacred Scripture, it cannot ultimately be contrary to the claims of legitimate science since God is the author of both and He cannot deceive based on what is knowable of Him on the basis of Sacred Scripture and Sacred Tradition. That does not mean that man has derived the proper interpretation of scripture in all circumstances or arrived at the final truth about nature as discovered by empirical science. It is necessary to constantly seek the truth in both faith (Divine Revelation) and science. Both scripture and contemporary astrophysics agree that the universe began to exist where nothing existed before that is, the universe had no material antecedent. This is completely compatible with the data from the philosophy of nature and metaphysics which holds that the material realm cannot be its own first cause. **Thus, relevant science, philosophy and revealed theology are all in agreement about the origin of the universe.** The importance of this cross-disciplinary concordance cannot be underestimated. It represents an extremely strong inferential (inductive) case that the immaterial first cause of the material universe is the God of the Bible, the God described in Christianity given the attributes which are given Him by Divine Revelation and which are demonstrated through modern science properly interpreted and without a-priori commitment to philosophical naturalism (materialism).

8.) Radical Darwinists who embrace philosophical naturalism are hopelessly misguided in thinking that Darwin in the words of Richard Dawkins "made it possible to be an intellectually fulfilled atheist." If Darwinists are serious about limiting science to methodological naturalism they must admit that it would never be possible to say anything about the existence or non-existence of God by studying science. If radical Darwinists wish to establish the non-existence of God while adhering to methodological naturalism, they will have to do so by philosophical not scientific means. In that arena they have few if any tools with which to make the case. Philosophy establishes beyond reasonable doubt that God exists and this has been true since Aristotle correctly reasoned that the material realm required an immaterial first cause, and certainly St. Thomas Aquinas no doubt the greatest philosopher of all time, developed his "5" Ways. Sacred Scripture indicates that a personal God exits.[113] Theology

---

[113] Careful analysis indicates that the factual matters with which scripture deals are in fact verifiable beyond reasonable doubt and as such, scripture should be accepted as the inspired word of God that is, a legitimate and true communication from God to His creatures.

provides a rich understanding of God's attributes and expectations. Properly understood and without a-priori acceptance of philosophical naturalism, science can also provide an understanding of God's creative activity and intentions through an understanding of His design of the universe and life itself. This becomes increasingly more not less possible in light of more recent scientific discoveries.

Selected Bibliography

Alberts, B. "The cell as a collection of protein machines: Preparing the next generation of molecular biologists. *Cell* 92: 291-294, 1998.

Arthur, W. The Origin of Animal Body Plans. Cambridge: Cambridge University Press, 1997.

Axe, D. D. "Biological Function Places Unexpectedly Tight Constraints on Protein Sequences." *Journal of Molecular Biology* 301 (3): 585-596, 2000.

Aizawa, S. I. "Flagellar assembly in Salmonella typhimurium." *Molecular biology* 19: 1-5, 1996.

Bacon, F. [1605]. The Advancement of Learning. Oxford: Clarendon Press, 1868.

Baron, E. et al., "Type IIP Supernovae as Cosmological Probes: A Spectral-Fitting Expanding Atmosphere Model Distance to SN 1999em," *Astrophysical Journal Letters* 616 (2004): L91-L94.

Barrow, John D. The Constants of Nature: From Alpha to Omega- The Numbers that Encode the Deepest Secrets of the Universe. New York: Pantheon Books, 2002.

Barrow, John D. and Frank J. Tippler. The Anthropic Cosmological Principle, New York: Oxford University Press, 1988, paperback edition.

Blakeley M. P. et al., "The 15° K Neutron Structure of Saccaride-free Concanavalin A," *Proceedings of the National Academy of Sciences* 101 (2004): 16405-10.

Behe, Michael J. Darwin's Black Box: The Biochemical Challenge to Evolution, New York: The Free Press, 1996.

Bowie, J. U., J.F. Reidhaar-Olson, W. A. Lim, and R. T. Sauer. "Deciphering the message in protein sequences: Tolerance to amino acid substitution." *Science* 247: 1306-1310, 1990.

Carter, Brandon. "Large Number Coincidences and the Anthropic Principle in Cosmology," *Proceedings of the International Astronomical Union Symposium, No. 63:* Confrontation of Cosmological Theories with Observational Data, ed. M. S. Longair, Dordrecht-Holland/Boston, U.S.A.: D. Reidel, 1974.

Carter, Brandon. "The Anthropic Principle and Its Implications for Biological Evolution," *Philosophical Transactions of the Royal Society, Series A* 370 (1983): 347-60.

Chen, Jun-Yung et al., "An Early Cambrian Craniate-like Chordate," *Nature* 402 (1999):

518-22.

Conaway, Joan Weliky and Ronald C. Conaway, "Light at the End of the Channel," *Science* 288 (2000): 632-33.

Craig, William Lane Reasonable Faith: Christian Truth and Apologetics. Wheaton, Ill: Crossway Books, 1984.

Cramer, Patrick et al., "Architecture of RNA Polymerase II and Implications for the Transcription Mechanism," *Science* 288 (2000): 640-49.

Cronin, J. R., S. Pizzarello, and D. P. Cruikshank, "Organic Matter in Carbonaceous Chrondrites, Planetary Satellites, Asteroids, and Comets," in *Meteorites and the Early Solar System*, eds. John F. Kerridge and Mildred Shapley Matthews (Tuscon: University of Arizona Press, 1988): 819-57

Darwin, Charles. On the Origin of Species. London: John Murray, 1859.

Davies, Paul The Fifth Miracle: The Search for the Origin and Meaning of Life. New York: Simon & Schuster, 1999.

Dawkins, Richard. River Out of Eden, New York: Harper Collins, 1995.

Dawkins, Richard. The Blind Watchmaker: Why the Evidence of Evolution Reveals a Universe without Design. New York: W.W. Norton & Company, 1996.

Deich J. et al., "Visualization of the Movement of Single Histidine Kinase Molecules in Live Caulobacter Cells," *Proceedings of the National Academy of Sciences, USA* 101 (2004): 15921-26.

Dembski, William A. No Free Lunch: Why Specified Complexity cannot be purchased without Intelligence, Lanham, Md. Rowman and Littlefield Publishers Inc., 2002.

Dembski, William A. The Design Inference: Eliminating Chance Through Small Probabilities. Cambridge: Cambridge University Press, 1998.

Dembski, William A. Uncommon Dissent: Intellectuals who Find Darwinism Unconvincing. Wilmington Delaware: ISI Books, 2004.

Dembski, William A. and Michael Ruse. Debating Design: From Darwin to DNA. Cambridge: Cambridge University Press, 2004.

Dennett, Daniel. Darwin's Dangerous Idea. New York: Simon and Schuster, 1995.

Denton, Michael. Evolution: A Theory in Crisis, London: Burnett Books Limited, 1985, U.S. distributor: Woodbine House, Bethesda, Md.

Denton, Michael J. Nature's Destiny: How the Laws of Biology Reveal Purpose in the Universe. New York: The Free Press, a Division of Simon and Shuster, 1998.

De Rosier, D. J. "The turn of the screw: The bacterial flagellar motor." *Cell* 93: 17-20, 1998.

D' Hondt, Steven et al., "Distributions of Microbial Activities in Deep Subseafloor Sediments," *Science* 306 (2004): 2216-21.

Dietl, Gregory P., Gregory S. Herbert, and Geerat J. Vermeij, "Reduced Competition and Altered Feeding Behavior Among Marine Snails After a Mass Extinction," *Science* 306 (2004): 2229-31.

Dorus, Steve et al., "Accelerated Evolution of Nervous System Genes in the Origin of Homo sapiens," *Cell* 119 (2004), 1027-1040.

Dupressoir, Anne et al., "Syncytin-A and Syncytin-B, Two Fusogenic Placenta-Specific Murine Envelope Genes of Retroviral Origin Conserved in Muridae," *Proceedings of the National Academy of Sciences, USA* 102 (2005): 725-30.

Dutcher, S. K. "Flagellar assembly in two hundred and fifty easy-to-follow steps." *Trends in Genetics* 11: 398-404, 1995.

Eldredge, Niles. The Triumph of Evolution and the Failure of Creationism, New York: W. H. Freeman, 2000.

Elsila, Jamie E. et al., "Mechanisms of Amino Acid Formation in Interstellar Ice Analogs," *Astrophysical Journal* 660 (2007): 911-18.

Emory, Nathan J. and Nicola S. Clayton, "The Mentality of Crows: Convergent Evolution of Intelligence in Corvids and Apes," *Science* 306 (2004): 1903-07.

Espinoza, Robert E. et al., "Recurrent Evolution of Herbivory in Small, Cold-Climate Lizards: Breaking the Ecophysiological Rules of Reptilian Herbivory," *Proceedings of the National Academy of Sciences,* USA 101 (2004): 16819-24.

Eyre-Walker, Adam and Peter D. Keightley, "High Genomic Deleterious Mutation Rates in Hominids," *Nature* 397 (1999).

Faulkner, A. J. et al., "PSR J1756-2251: A New Relativistic Double Neutron Star System," *Astrophysical Journal Letters* 618 (2005): L119-L122.

Frese Raoul N. et al., "The Long-Range Organization of a Native Photosynthetic Membrane," *Proceedings of the National Academy of Sciences, USA* 101 (2004): 17994-99.

Fukugita, Masataka and P. J. E. Peebles, "The Cosmic Energy Inventory," *Astrophysical Journal* 616 (2004): 643-68.

Futuyma, Douglas J. Evolutionary Biology. Third edition. Sinauer, Sunderland: MA. 1988.

Gonzalez, Guillermo and Jay W. Richards. The Privileged Planet: How Our Place in the Cosmos is Designed for Discovery. Washington, D.C. Regnery, 2004.

Grossman, Lawrence I. et al., "Accelerated Evolution of the Electron Transport Chain in Anthropoid Primates," *Trends in Genetics* 20 (2004): 578-85.

Haldane, J.B.S. *Rationalist Annual* 148, 1928, pp. 3-10.

Hawking, Stephen. A Brief History of Time: From the Big Bang to Black Holes. New York: Bantam Books, 1988 (paper back edition).

Heintz, Nathaniel."Survival by Self-Digestion," *Nature* 432 (2004): 963.

Hiratani, Ichiro et al., "Differentiation-Induced Replication-Timing Changes are Restricted to AT-Rich/Long Interspersed Nuclear Element (LINE)-Rich Isochors," *Proceedings of the National Academy of Sciences*, USA 101 (2004): 16861-66.

Huemmerich, Daniel et al., "Novel Assembly Properties of Recombinant Spider Dragline Silk Proteins," *Current Biology* 14 (2004): 2070-74.

Jaki, Stanley L. The Bible and Science, Front Royal Virginia: Christendom Press, 1996.

Jaki, Stanley L. The Savior of Science, Grand Rapids, Michigan: William B. Eerdmans Publishing Co. 2000.

Johnson, Philip E. Darwin on Trial, Downers Grove Illinois: InterVarsity Press, 1993.

Johnson, Phillip E. The Wedge of Truth: Splitting the Foundations of Naturalism. Downers Grove, Ill: InterVarsity Press, 2000.

Kazmirski, Steven L et al., "Structural Analysis of the Inactive State of the Escherichia coli DNA Polymerase Clamp-Loader Complex," *Proceedings of the National Academy of Sciences,* USA 101 (2004): 16750-55.

Keightley, Peter D. et al., "Evidence for Widespread Degradation of Gene Control Regions in Hominid Genomes," *PLOS Biology* (February 2005) e42.

Kerr, Richard A. "Did Jupiter and Saturn Team Up to Pummel the Inner Solar System?

Report from the November 8-12, 2004 Meeting of the Division for Planetary Sciences at Louisville, Kentucky," *Science* 306 (2004): 1676.

Körtzinger, Arne et al., "The Ocean Takes a Deep Breath," *Science* 306 (2004): 1337.

Kouzminova, Elena A. et al., "RecA-Dependent Mutants in Escherichia coli Reveal Strategies to Avoid Chromosomal Fragmentation," *Proceedings of the National Academy of Sciences*, USA 101 (2004): 16262-67.

Krauss, Lawrence M. "The End of the Age Problem and the Case for a Cosmological Constant Revisited," *Astrophysical Journal* 501 (1998): 461-66.

Kreeft, Peter, Socratic Logic: A Logic Text Using Socratic Method, Platonic Questions, and Aristotelian Principles. South Bend, Indiana: St. Augustine Press, 2004.

Krug, A. Z. and M. E. Patzkowsky, "Rapid Recovery from the Late Ordovician Mass Extinction," *Publication of the National Academy of Sciences* 101 (2004): 17605-10.

Kuan, Yi-Jehng et al., "Interstellar Glycine," *Astrophysical Journal* 593 (2003): 848-67.

Kuma, Akiko et al., "The Role of Autophagy During the Early Neonatal Starvation Period," *Nature* 432 (2004): 1032-36.

Kvenvolden, Keith A., James G. Lawless, and Cyril Ponnamperuma, "Nonprotein Amino Acids in the Murchison Meteorite," *Proceedings of the National Academy of Sciences, USA* 68 (1971): 486-90

Lennon, Brett W. et al., "Twists in Catalysis: Alternating Conformations of *Escherichia Coli* Thioredoxin Reductase," *Science* 289 (2000): 1190-94.

Leslie, John. Universes. London: Routledge, 1989.

Lipton, P. Inference to the Best Explanation. New York: Routledge, 1991.

Liu, Wan-chun et al., "Juvenile Zebra Finches Can Use Multiple Strategies to Learn the Same Song," *Proceedings of the National Academy of Sciences, USA* 101 (2004): 18177-82.

Maria, Steven F. et al., "Organic Aerosol Growth Mechanisms and Their Climate-Forcing Implications," *Science* 306 (2004): 1921-24.

Mayr, Ernst. What Evolution is. New York: Basic Books, 2001.

Miller, Kenneth R. Finding Darwin's God: A Scientist's Search for Common Ground Between God and Evolution. New York: Harper Collins (Perennial Edition),

2002.

Mojzsis S. J. et al., "Evidence for Life on Earth before 3,800 Million Years Ago," *Nature* 384 (1996), 55-59.

Moreland, J.P. Scaling the Secular City, Grand Rapids Mi.: Baker Book House, 1987.

Naselsky, Pavel D. et al., "Primordial Magnetic Field and Non-Gaussianity of the One-Year Wilkinson Microwave Anisotropy Probe Data," *Astrophysical Journal* 615 (2004): 45-54.

Oparin, A. I. The Origin of Life, (Russian Proiskhozdenic Zhizny), Moscow: Moskovski Rabochii, 1924.

Ostfield, Richard S. and Felicia Keesing, "Oh the Locusts Sang, Then They Dropped Dead," *Science* 306 (2004): 1488-89.

Ovcharenko, Ivan et al., "Evolution and Functional Classification of Vertebrate Gene Deserts," *Genome Research* 15 (2005): online preprint.

Overman, Dean. A case Against Accident and Self-Organization. New York: Rowman & Littlefield Pub., 1997.

Parsons, Paul "Dusting Off Panspermia," *Nature* 383(1996), 221-22.

Pennock, R. Tower of Babel: The Evidence against the New Creationism. Cambridge, MA: MIT Press, 1999.

Pope Pius XII, *Humani Generis,* Rome: The Holy See, 1950.

Rana, Fazale and Hugh Ross. Origins of Life: Biblical and Evolutionary Models Face Off. Colorado Springs, Colorado: NavPress, 2004.

Rana, Fazale R. "Convergence: Evidence for a Single Creator" *Facts for Faith*, issue 4, 2000.

Reichenbach, Bruce, R. The Cosmological Argument: A Reassessment. Springfield, Ill.: Charles C. Thomas Publishers, 1972.

Ribeiro, Guy M. and Sebastian Bonhoeffer, "Production of Resistant HIV Mutants During Antiretroviral Therapy," *Proceedings of the National Academy of Sciences, USA* 97 (2000): 7681-86.

Rosing, Minik T. "13C-Depleted Carbon Microparticles in 3700-Ma Sea-Floor Sedimentary Rocks from West Greenland," *Science* 283 (1999), 674-76.

Ross, Hugh, PhD. Beyond the Cosmos: What Recent Discoveries in Astronomy and
    Physics Reveal about the Nature of God. Colorado Springs, Colorado: NavPress,
    1996.

Ross, Hugh, PhD. The Genesis Question: Scientific Advances and the Accuracy of
    Genesis. Colorado Springs, Colorado: NavPress, 1998.

Ruddiman, William F., Stephen J. Vavrus, and John E. Kutzbach, "A Test of the Over
    due-Glaciation Hypothesis," *Quaternary Science Reviews* 24 (2005): 1-10.

Sala, Monica and Simon Wain-Hobson, "Are RNA Viruses Adapting or Merely
    Changing?" Journal of Molecular Education 51 (2000): 12-20.

Samatey Fadel A. et al., "Structure of the Bacterial Flagellar Hook and Implication for
    the Molecular Universal Joint Mechanism," *Nature* 431 (2004): 1062-68.

Schidlowski, Manfred. "A 3,800-Million-Year Isotopic Record of Life from Carbon in
    Sedimentary Rocks," Nature 333 (1988).

Schidlowski, Manfred. "Carbon Isotopes as Biogeochemical Recorders of Life over 3.8
    Ga of Earth History: Evolution of a Concept," *Precambrian Research* 106 (2001):
    117-34.

Schopf, J. William. "Microfossils of the Early Archean Apex Chert: New Evidence of the
    Antiquity of Life," *Science* 260 (1993), 640-46.

Seelert, Holger et al., "Proton-Powered Turbine of a Plant," *Nature* 405 (2000): 418-19.

Semaw, Selishi et al.,"Early Pliocene Hominids from Gona, Ethiopia," *Nature*
    433 (2005): 301-05.

Shu, D. G. et al., "An Early Cambrian Tunicate from China," *Nature* 411 (2001): 472-3.

Shu, D. G. et al., "Lower Cambrian Vertebrates from South China," *Nature* 402 (1999):
    42-46.

Shu, D. G. et al., "Primitive Deuterostomes from the Chengjiang Lagerstatte (Lower
    Cambrian, China)," *Nature* 414 (2001): 419-24.

Snyder, L. E. et al., "A Rigorous Attempt to Verify Interstellar Glycine," *Astrophysical*
    Journal 619 (2005): 914-30

Sober, Elliot. The Philosophy of Biology, 2nd. Ed. San Francisco: Westview Press, 2000.

Soong Ricky K. et al., "Powering an Inorganic Nanodevice with a Biomolecular Motor,"
    Science 290 (2000), 1555-58.

Swinburne, Richard. "Argument from the Fine-Tuning of the Universe," Physical Cosmology and Philosophy, ed. John Leslie New York: Macmillan, 1991, 165.

Thaxton, Charles B., Walter L. Bradley and Roger L. Olsen. The Mystery of Life's Origin: Reassessing Current Theories. Dallas TX: Lewis and Stanley, 1984.

The Genesis Debate: Three Views on the Days of Creation. Edited by David G. Hagopian, Mission Viejo, California: Crux Press, Inc. 2001.

Valiyaveetil, Francis I. et al., "Glycine as a D-Amino Acid Surrogate in the K+- Selectivity Filter," Proceedings of the National Academy of Sciences, USA 101 (2004): 17045-49.

Van der Wielen, Paul W. J. J. et al., "The Enigma of Prokaryotic Life in Deep Hyper saline Anoxic Basins," Science 307 (2005): 121-23.

Van Loon, A. J. "The Needless Search for Extraterrestrial Fossils on Earth," Earth-Science Reviews 68 (2005): 335-46.

Verlet, J. R. R. et al., "Observation of Large Water-Cluster Anions with Surface-Bound Excess Electrons," Science 307 (2005): 93-96.

Wallace, William A. The Elements of Philosophy: A Compendium for Philosophers and Theologians. New York: Alba House, 1977.

Wallace, William A. The Modeling of Nature: Philosophy of Science and Philosophy of Nature in Synthesis. Washington D.C.: Catholic University of America Press, 1996.

Ward, Peter D. and Donald Brownlee, Rare Earth: Why Complex Life Is Uncommon in the Universe. New York: Springer-Verlag, 2000.

Waltham, Dave. "Anthropic Selection for the Moon's Mass," Astrobiology 4 (2004): 460-68.

Welte, Michael A. "Bidirectional Transport along Microtubules," Current Biology 14 (2004): R525-R537.

Weinberg, Steven. The First Three Minutes: A Modern View of the Origin of the Universe. New York: Basic Books, 1988, (paper back edition).

Wenger, Oliver S. et al., "Electron Tunneling Through Organic Molecules in Frozen Glasses," Science 307 (2005): 99-102.

Westall, Frances et al., "Early Archean Fossil Bacteria and Biofilms in Hydrothermally-

Influenced Sediments from Barberton Greenstone Belt, South Africa," *Precambrian Research* 106 (2001): 93-116.

Yang, Louie H. "Periodical Cicadas as Resource Pulses in North American Forests," 306 *Science* (2004): 1565-67.

Yockey, Hubert. Information Theory and Molecular Biology. Cambridge: Cambridge University Press, 1992.

Yockey, H. P. "A Calculation of the Probability of Spontaneous Biogenesis by Information Theory." *Journal of Theoretical Biology* 67:377-398, 1978.

Yonekura, K., S. Maki, D. G. Morgan, D. J. DeRosier, F. Vonderviszt, K. Imada, and K. Namba. "The bacterial flagellar cap as the rotary promoter of flagellin self-assembly." *Science* 290:2148-2152, 2000.

Zahale, K. J. and N. H. Sleep, "Impacts and the Early Evolution of Life" Comets and the Origin and Evolution of Life. Edited by Paul J. Thomas, Christopher F. Chyba, and Christopher P. McKay (New York: Springer-Verlag, 1997), 175-208.

Zhao, Yong et al., "Reversed Voltage-Dependent Gating of a Bacterial Sodium Channel with Proline Substitutions in the S6 Transmembrane Segment," *Proceedings of the National Academy of Sciences, USA* 101 (2004): early edition.

76